Illustrated Science Dictionary

Elise Masters

Viva Books

New Delhi | Mumbai | Chennai | Kolkata | Bangalore | Hyderabad | Kochi

First Indian Edition 2008

VIVA BOOKS PRIVATE LIMITED

- 4737/23 Ansari Road, Daryaganj, New Delhi 110 002
  E-mail: vivadelhi@vivagroupindia.net, Tel. 42242200
- Plot No. 76, Service Industries, Shirvane, Sector 1, Nerul, Navi Mumbai 400 706
  E-mail: vivamumbai@vivagroupindia.net, Tel. 27721273, 27721274
- Jamals Fazal Chambers, 26 Greams Road, Chennai 600 006
  E-mail: vivachennai@vivagroupindia.net, Tel. 28290304, 28294241
- B-103, Jindal Towers, 21/1A/3 Darga Road, Kolkata 700 017
  E-mail: vivakolkata@vivagroupindia.net, Tel. 22836381, 22816713
- 7, Sovereign Park Apartments, 56-58, K. R. Road, Basavanagudi, Bangalore 560 004
  E-mail: vivabangalore@vivagroupindia.net, Tel.26607409, 26607410
- 101-102 Mughal Marc Apartments, 3-4-637 to 641, Narayanguda, Hyderabad 500 029
  E-mail: vivahyderabad@vivagroupindia.net, Tel. 27564481, 27564482
- Beevi Towers, First Floor, SRM Road, Kaloor, Kochi - 682 018, Kerala
  E-mail: vivakochi@vivagroupindia.net

Visit us at www.vivagroupindia.com

Published by arrangement with
Pascal Educational Services
655 Parramatta Road
Leichhardt NSW 2040
Australia

Written by Katy Pike
Pictures by Luke Jurevicius
Series Publisher Katy Pike
Editor: Garda Turner
Design and layout by Modern Art Production Group

ISBN-10: 81-309-0503-5
ISBN-13: 978-81-309-0503-7

This edition is for sale in India, Nepal, Sri Lanka and Maldives only. Not for export elsewhere.

Published by Vinod Vasishtha for Viva Books Private Limited,
4737/23 Ansari Road, Daryaganj, New Delhi 110 002.

Printed and bound at Viba Press Pvt. Ltd., New Delhi-20

Contents

A 1
B 6
C 12
D 29
E 35
F 48
G 56
H 61
I 65
J 68
K 69
L 71
M 76
N 88
O 90
P 93
R 104
S 114
T 132
U 138
V 140
W 144
X 150
Y 151

Introduction

This dictionary has been written for school students. It is suitable for both primary and high-school classes. It contains an alphabetical list of common words and phrases used in the school science textbooks. Coloured, labelled diagrams and illustrations, as well as numerous examples, have been included to assist with understanding and learning.

Words in *italics* indicate where more information is included elsewhere in the book. Other related material is listed at the end of each entry in **bold** print.

A

absorption

See **heat absorption**.

acceleration

occurs when an object increases its *speed*. When the object slows down, it decelerates. Acceleration and deceleration are measured in metres per second per second (m/s^2).

Acceleration:

This plane is increasing its speed as it takes off.

Deceleration:

This plane is decreasing its speed as it comes to a stop.

acids

form an important group of chemical *compounds*. They can be detected using *indicators* (e.g. litmus and universal indicator). See **pH**.

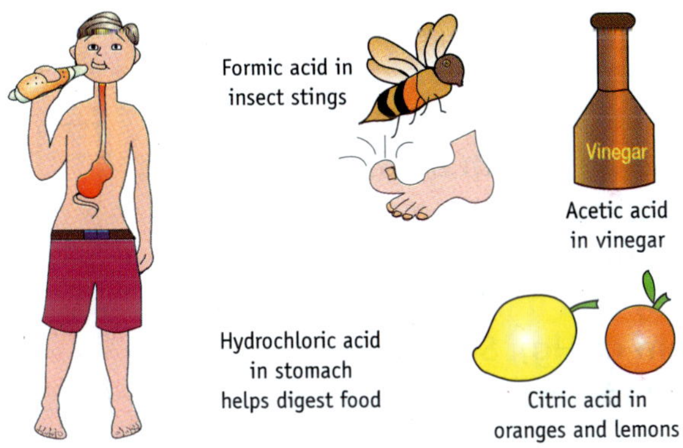

adaptation

is a feature or behaviour that helps *living things* to survive. See **survival features; reproductive adaptations**.

Desert lizard: Some desert lizards stay cool by burrowing.

Human face: Humans sweat when hot; *evaporation* of the sweat keeps the body cool.

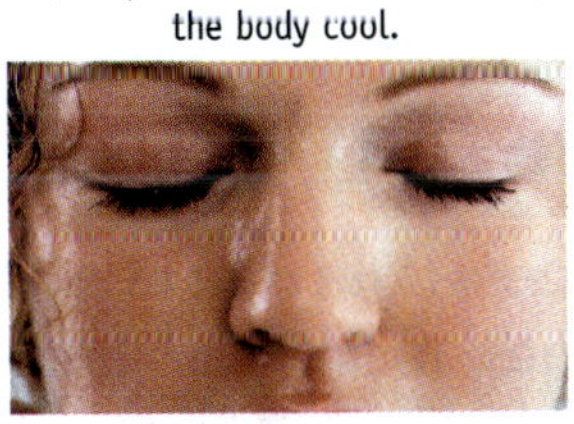

Polar bear: A polar bear has thick, layered fur, which holds in body heat.

Cactus: A cactus has a fleshy *stem* that stores water and spines that protect it from predators. See **predation**.

A

air

air

is a *mixture* of *gas*es that extends about 50 km around the *Earth*. Important gases are nitrogen (78%), *oxygen* (21%) and *carbon dioxide* (0.03%).

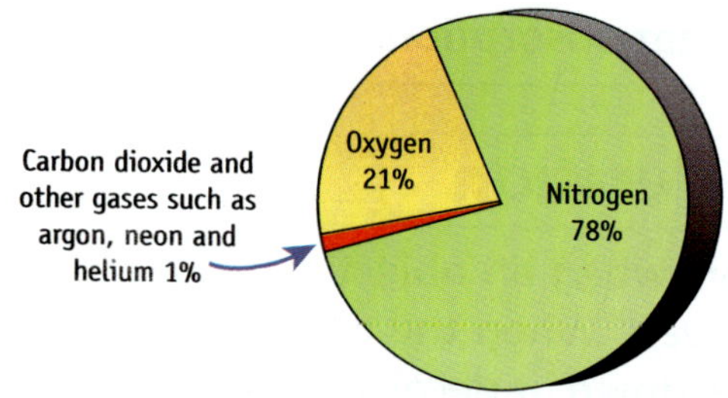

air currents

See **global air currents.**

air pressure

or atmospheric pressure is the *pressure* of the *air* on its surroundings. Air pressure is measured with a *barometer*. It is measured in kilopascals (kPa) and hectopascals (hPa), millibars (mB) and millimetres of mercury (mm of Hg). See **aneroid barometer; high pressure system; low pressure system.**

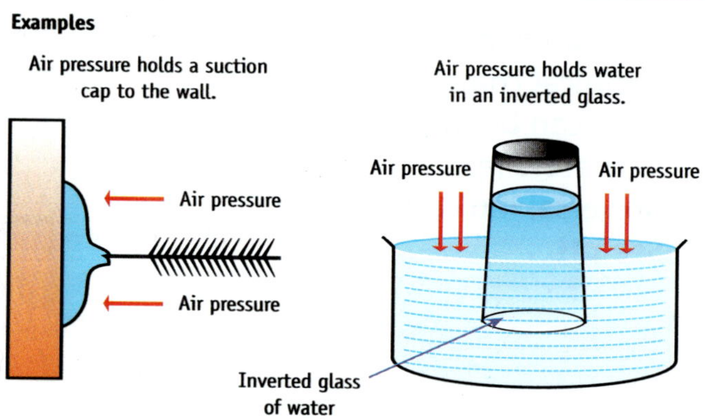

air pressure and weather

See **high pressure system; low pressure system.**

air resistance

occurs when *air* slows the movement of an object. Planes, racing cars and birds are streamlined to reduce air resistance. Parachutes use air resistance to slow movement. See **friction reduction.**

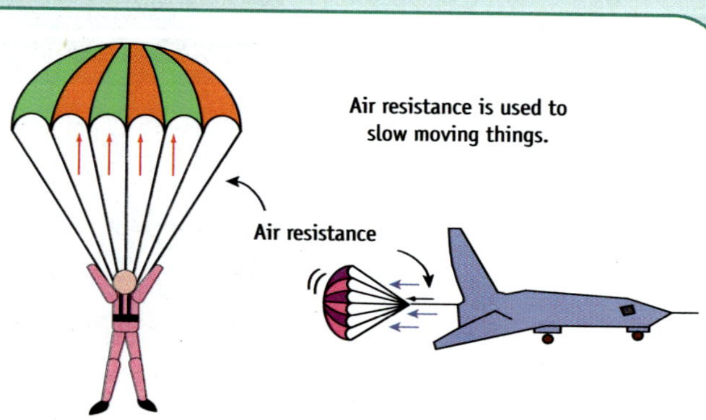

alloys

are *mixtures* of different *metals*. Making metals into alloys gives them different *properties* and makes them useful for certain jobs (e.g. stainless steel made of iron, chromium and nickel is stronger than iron and does not *rust*).

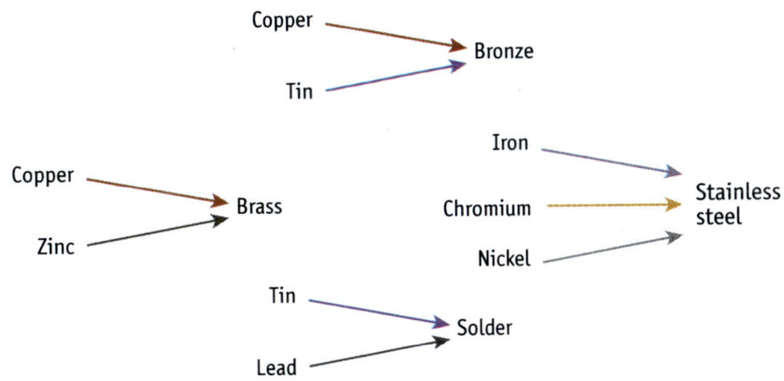

alternating current

is when the *current* or *charge* moves backwards and forwards. Many *generators* produce this type of current. Compare **direct current (DC)**.

anemometer

measures the *speed* of *wind*. Wind from any direction will make the cups on the anemometer spin. Anemometers may be connected to computers for accurate readings.

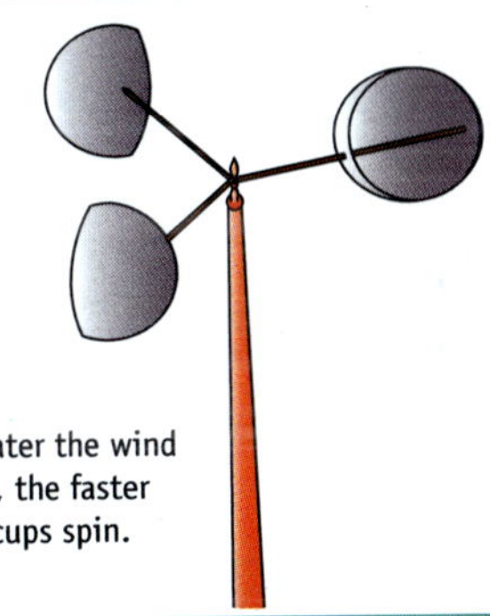

The greater the wind speed, the faster the cups spin.

aneroid barometer

measures *air pressure*. It contains a sealed box with most of its *air* removed. When the surrounding air pressure changes, the box becomes larger or smaller. These changes in the size of the box are recorded on a dial.

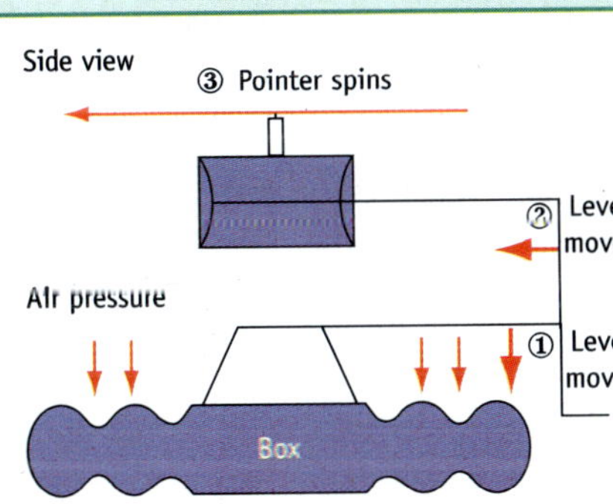

Change in air pressure is magnified by a lever system which moves a pointer over a dial.

A

animals

are *organisms* that obtain their *food* by eating *plants* or other animals. Generally they move about and react quickly to changes. Their *cells* have no wall.
See **vertebrates; invertebrates**.

aquatic

refers to anything living or growing in water (fresh or salt). Aquatic *environments* include seas, oceans, estuaries, rivers, lakes and rock pools.

asexual reproduction

is when new *organisms* are produced by the simple dividing of *cells*. They are copies of the parent. Compare **sexual reproduction**.

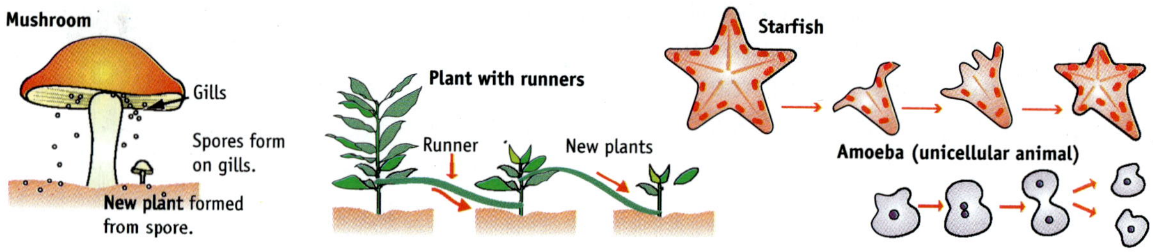

asteroids

are large rock fragments in a band *orbiting* the *Sun*. They are mainly between *Mars* and *Jupiter*.
See **Solar System**.

asthenosphere

is an area in the upper *mantle* of the *Earth* that is less rigid and slowly oozes. It allows the plates that make up the *lithosphere* to move.

atmosphere

is a blanket of *gas*es surrounding a *planet*. The *Earth's* atmosphere extends about 800 km. Most of the *air* is in the first 50 km. The atmosphere protects the Earth from *meteor*s and harmful rays (e.g. *ultraviolet rays*). The atmosphere is made up of layers. See **atmosphere, changes**.

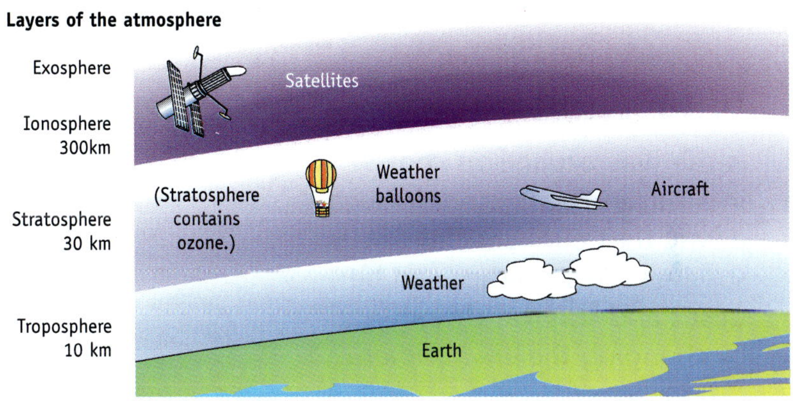

atmosphere, changes

have occurred over time. When the *Earth* was formed the atmosphere was mainly made up of *carbon dioxide*. Today, the atmosphere is mainly nitrogen and *oxygen*. The present atmosphere is maintained by the activities of *plants* and *animals*.

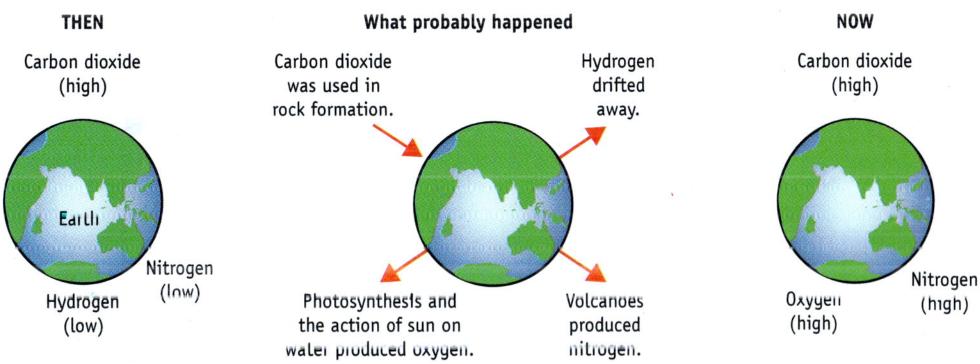

atoms

are extremely small particles that make up *matter*. An atom contains charged particles, positive protons and negative electrons. All the atoms in an *element* are the same. The atoms of one element are different from those of a different element (e.g. the atoms in iron are different from those in copper). See **charge**. Compare with **molecule**.

B

bacteria (bacterium — singular)

are microscopic, *unicellular organisms*. Some cause disease, but many are useful (e.g. they break down dead organisms and *recycle nutrients*).
See **microorganisms**.

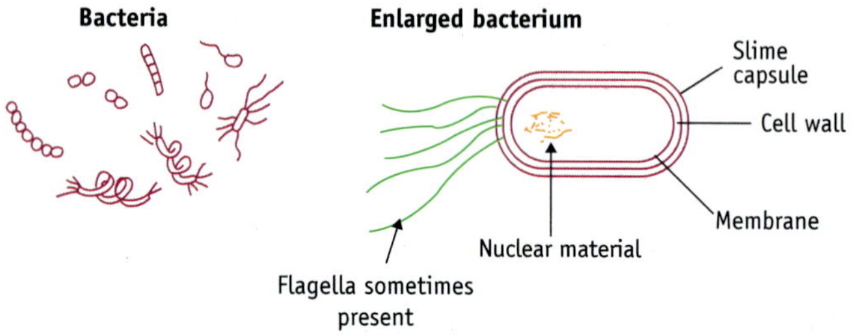

balance of nature

is the way *organisms* coexist in a natural *ecosystem* in such a way that each *species* survives. This balance can be upset and survival threatened by changing the *environment*.

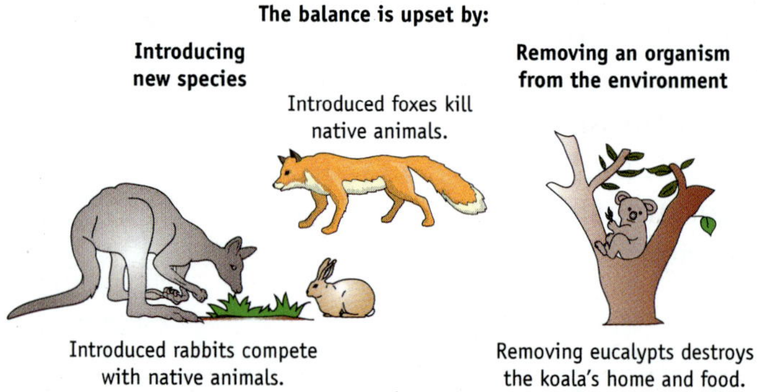

barometer

is an instrument for measuring *air* (atmospheric) pressure. A simple mercury barometer uses *air pressure* to hold up a column of mercury in a glass tube. If the air pressure is low, the column drops. See **aneroid barometer**.

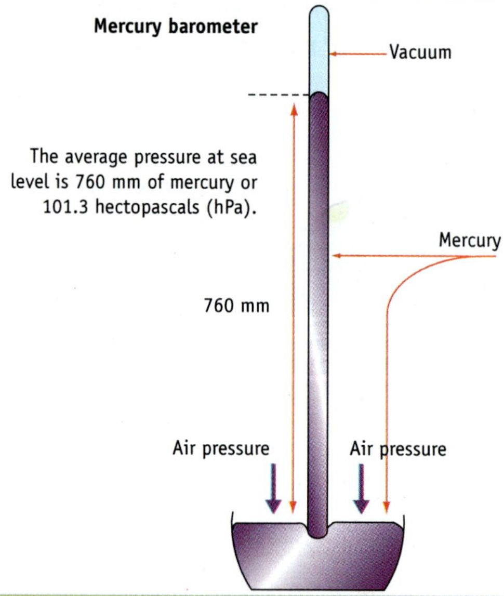

basalt

See **igneous rocks**.

batteries

are made of two or more *electric cells* (or dry cells). *Chemical reactions* in the cells create *electrical energy*. See **voltage**.

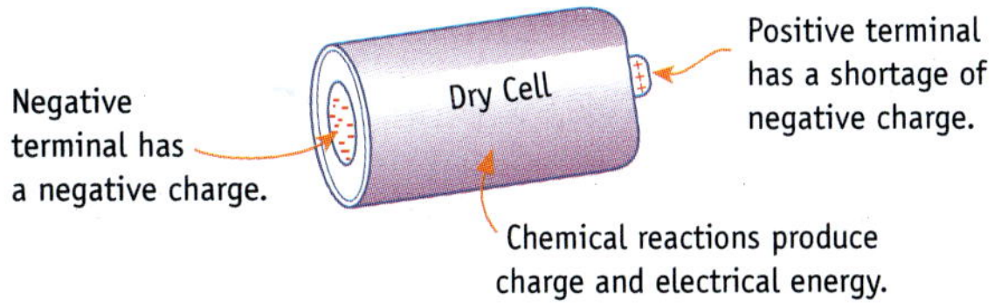

batteries in parallel

is when batteries are placed in parallel in a *circuit*. Their *voltage* or electrical push is not combined and *current* is not increased. Compare **batteries in series**.

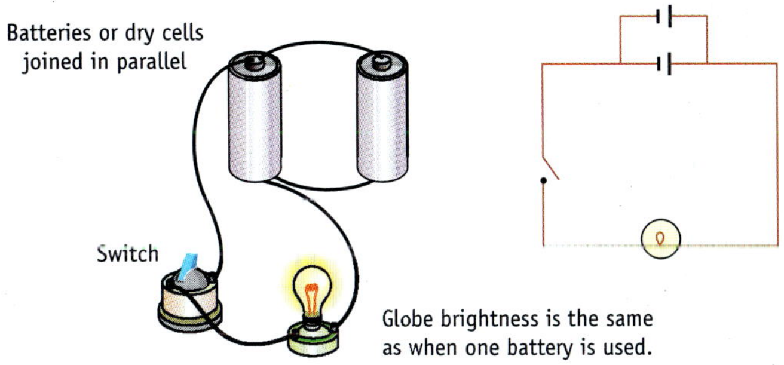

batteries in series

is when batteries are placed in a row in a *circuit*. Their *voltage* or electrical push is combined and more *current* can flow. Compare **batteries in parallel**.

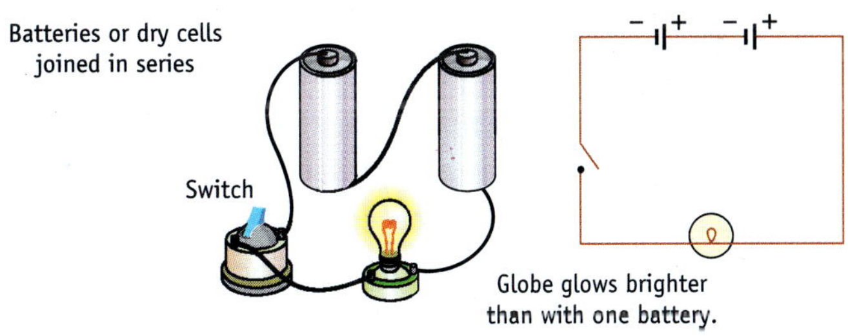

biodegradable

refers to a material that can be broken down or decomposed by *living things* such as bacteria. Decomposing the material makes it harmless so that it will not pollute. Some *plastic*s are now biodegradable. See **decomposers**; **soaps and detergents**.

B

biodiversity

is the variety of *living things* living on *Earth*. Variety is reduced by human activities that destroy *organisms*, their shelters and *food* (e.g. hunting, removing forests and setting up farms and cities). See **biodiversity**; **extinction**.

black hole

is the remains of a *star*. It has very powerful *gravity* so that nearby objects are sucked into it. Nothing can escape from a black hole, including *light*, thus making it black and invisible. Astronomers detect black holes by the effect of their gravity on nearby bodies in *space*.

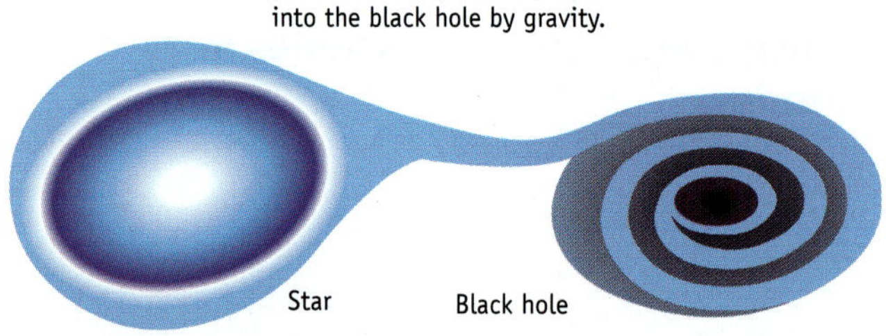

boiling

occurs when the particles in a *liquid* separate and form *gas* bubbles. The bubbles rise to the surface. A change of state occurs during boiling. Compare **melting**. See **boiling point**; **state of matter, change**.

boiling point

is the *temperature* at which a substance boils and turns to a *gas*. A *liquid* with strong *bonds* between its particles has a high boiling point (e.g. iron boils at 2750°C). A liquid with weak bonds has a low boiling point (e.g. water boils at 100°C). Compare **melting point**.

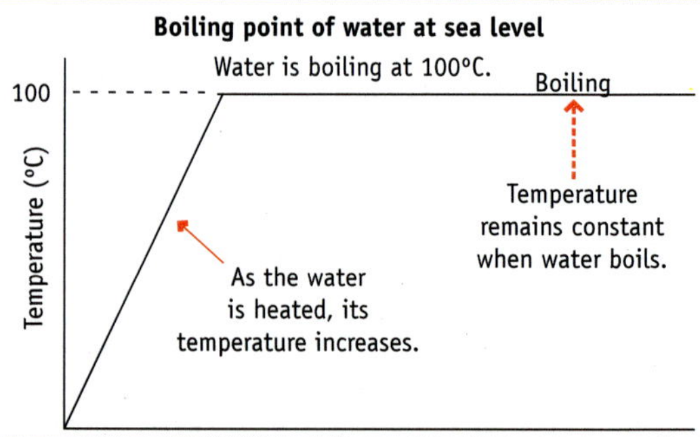

bonds

See **forces between particles in matter**.

breathing

is the process of taking in *oxygen* and giving out *carbon dioxide*. *Vertebrates* use lungs to breathe, fish use gills and insects have small pores down the sides of their bodies.

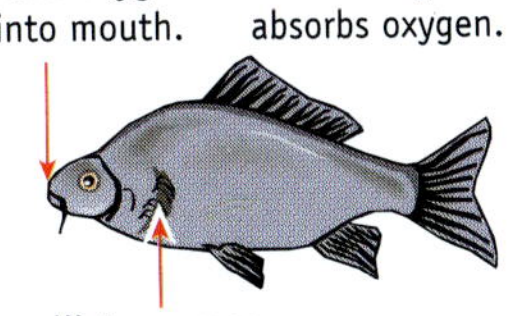

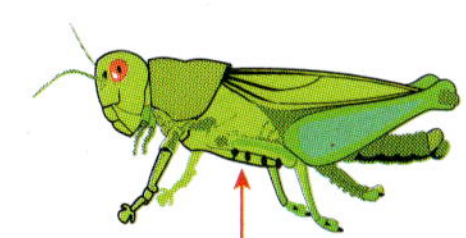

bridges

exert *forces* on the foundation and surrounding *rock*. See **foundations; building tension and compression**.

Pier bridges

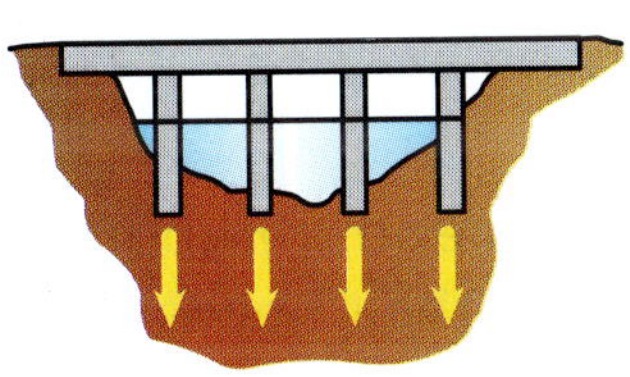

Pier bridges exert downward forces that are spread over the base rock. They can be built on most rock types.

Suspension bridges

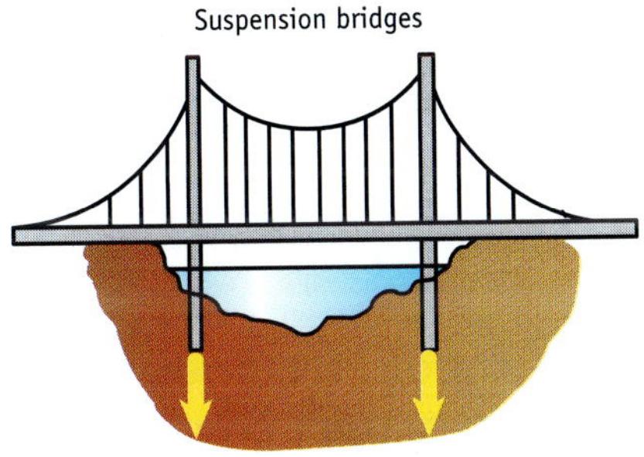

The pressure under the towers is great and the foundation rock needs to be strong.

Arch bridges

Arch bridges exert downward and sideways forces so the rock foundations need to be strong in all directions.

building and construction

form structures (e.g. buildings, *dams*, *bridges*, *tunnels*) that need to be strong to support their own *weight*. *Building materials* and designs are carefully selected so they can withstand *forces* of tension and *compression*. See **building tension and compression; foundations, material shaping**.

Beams are less likely to bend if they are supported by trusses.

1. Roofing trusses

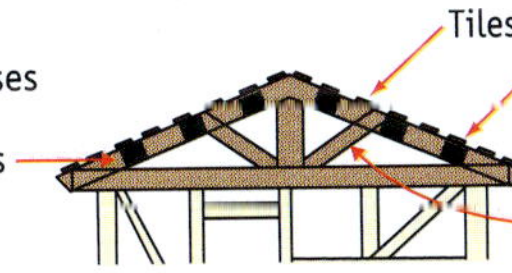

2. Bridge trusses

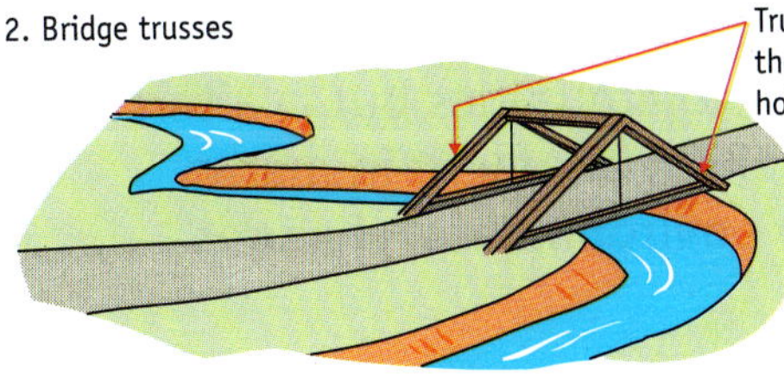

building materials

are selected for their *properties*. Stone, bricks and concrete have great compressive strength but crack when under tension. Both wood and steel are strong under *compression* and tension *forces*. See **building tension and compression; manufactured building materials; rock use; material shaping; material strengths.**

building tension and compression

affect all structures. All building structures are under *forces* of tension (pulling forces) and *compression* (pushing forces). See **building and construction; building materials; material shaping**.

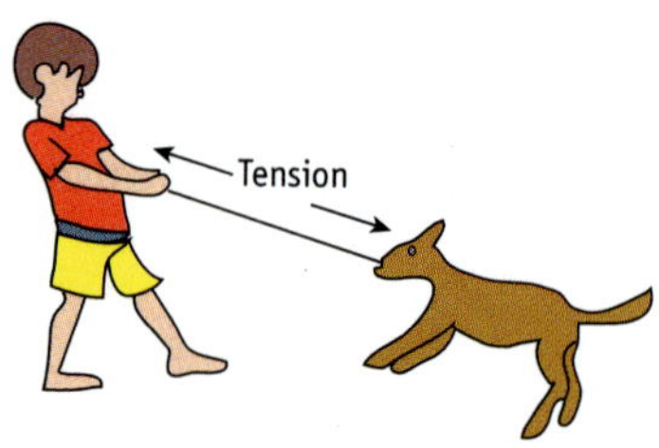

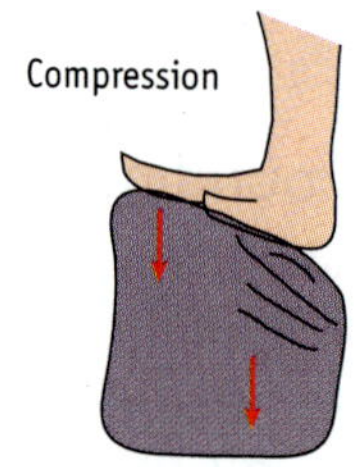

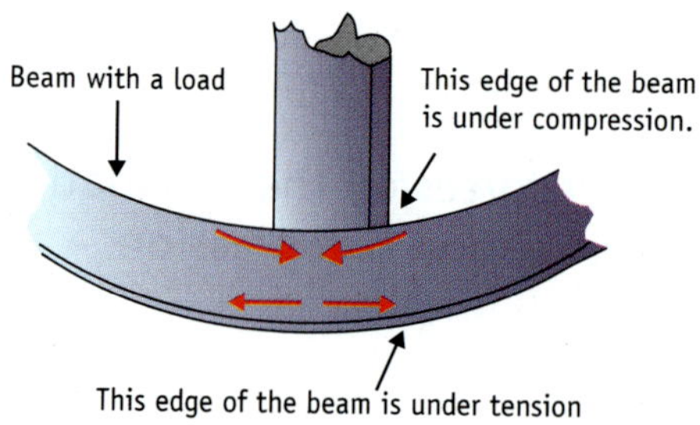

built environments

include towns, homes, shops, schools, factories, parks, gardens and *transport systems*. People modify and control their *environments* to suit their needs and so that they will work well and look attractive.

built environments, control systems

are set up to control conditions in human communities. Examples are traffic lights and law enforcement, air conditioning and *heat* insulation to control *temperature*, and sound proofing to control noise. See **insulator (heat)**.

built environments, impact

is the effect building has on the *environment*. Building is necessary for human communities but it alters the natural environment in many ways. See **environmental changes**.

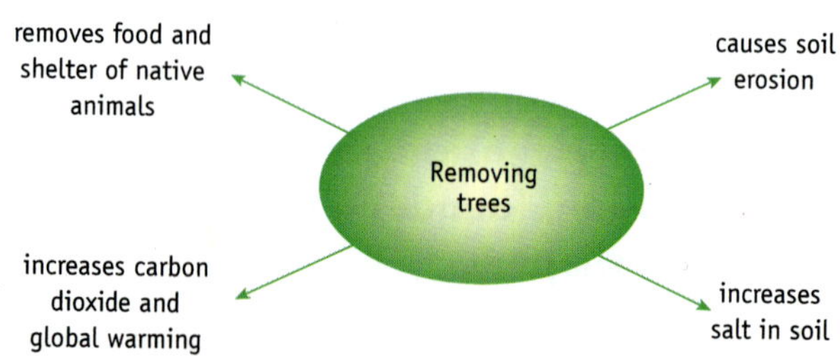

built environments, services

are needed by human communities to enable them to function. Services include water and *energy* (*electricity* and *gas*) supplies, transport, health and education services, *communication* systems (mail, newspapers, telephones) and *waste disposal* plants. See **energy sources**.

buoyancy

refers to an object that is hard to *sink*. It is said to be buoyant. An object is more buoyant in salt water than fresh water. See **float**; **sink**.

burning or combustion

is a *chemical reaction* that occurs when a substance is heated and combines with *oxygen* in the *air*. It produces *heat, light, sound* and new chemicals.

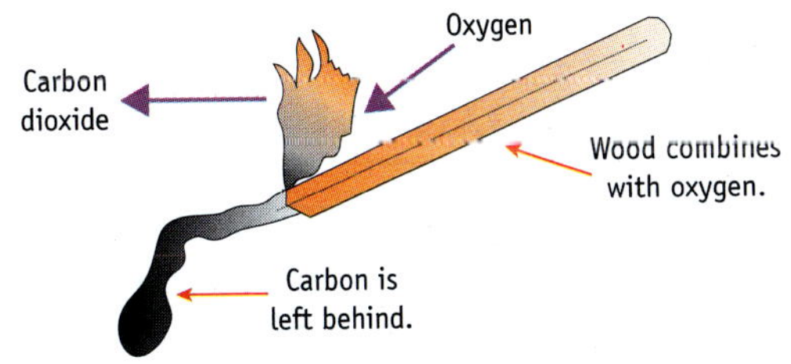

bushfires and floods

change *ecosystems*. They kill *plant*s and *animals* and destroy *food* and places where animals can hide from *predators*. They damage the *soil* and cause *erosion* by killing and removing plants.

cantilever

A large projecting structure or beam attached to one end. See **material shaping**.

carbohydrates

are *organic compounds* such as sugar, starch and glucose. They are used in *respiration* to provide *energy* for activities and *cell* functioning. Glucose is made by *plants* during *photosynthesis* and it may change to sugar or starch for storage. See **nutrients**.

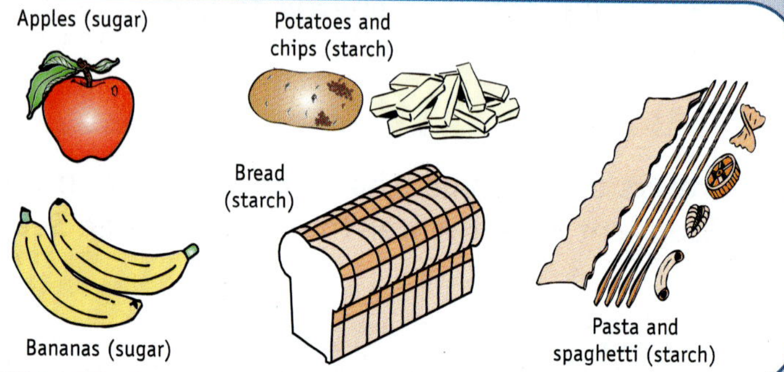

carbon dioxide (CO_2)

is a *gas* found in the *atmosphere*. *Plants* use it to make their *food* supply. Both plants and *animals* make carbon dioxide in *respiration*. *Burning* wood and *fossil fuels* releases carbon dioxide. See **greenhouse effect**.

carnivore

is an *animal* that eats the flesh of other animals (e.g. cat, lizard).

catalyst

is a chemical that speeds up reactions but does not take part in them. Some *chemical reactions* would take years without a catalyst. *Enzymes* are catalysts.

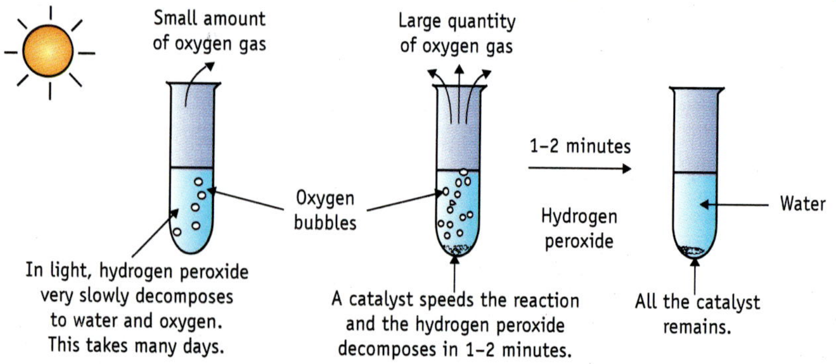

cell

See **cells in living things; electric cells.**

cells in living things

are small structures that make up the whole. Most are microscopic. The diagram shows details of cells seen with a *light* microscope. See **cells in living things, types**.

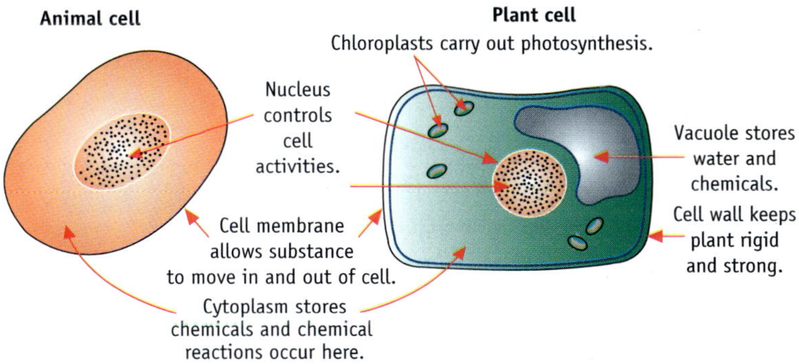

cells in living things, types

include many different forms. Each type has its own particular job. See **xylem.** For examples of cells in plants see **xylem, stomates.**

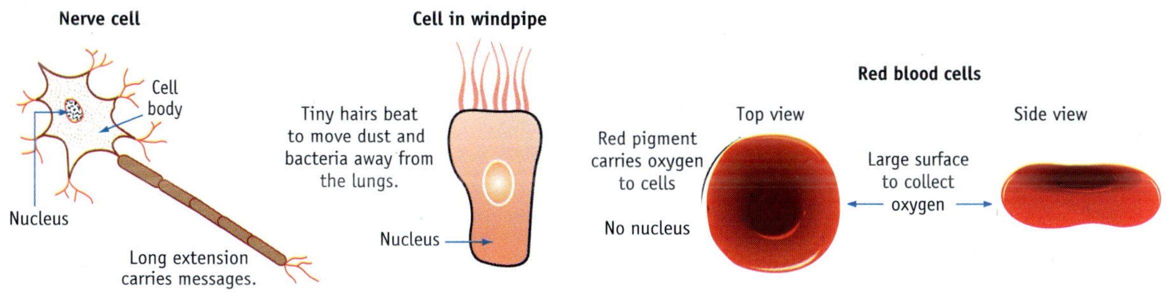

centrifuge

is a *machine* used to separate different *solids* or to separate *liquids* from solids by spinning. A washing machine spins to separate the water from the clothes; in laboratories centrifuges separate red blood *cells* from blood plasma.

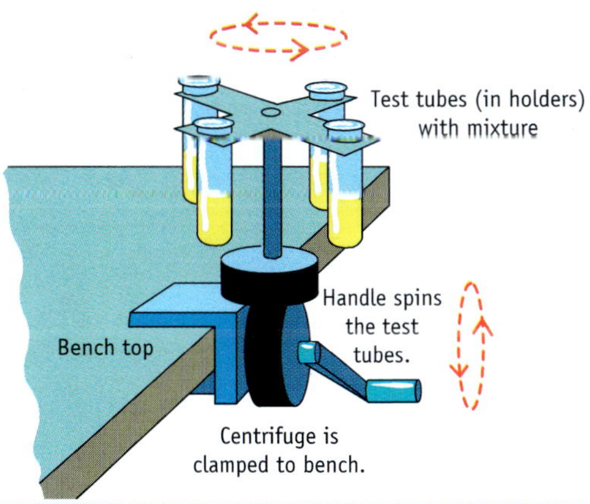

C

change of state

See **state of matter, change**.

charge

is positive or negative. *Matter* in an object contains positive and negative particles. If there are more negative than positive particles, the object has a negative charge. If there are more positive particles than negative, it has a positive charge. If positive and negative particles are equal, the object has no charge. See **static electricity; electrostatic force**.

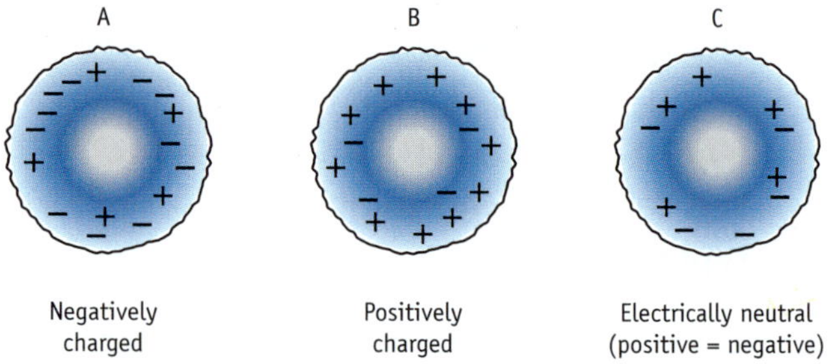

charging an object

may be done by rubbing it with another object. Rubbing causes the negative charge to move from one object to the other. The excess charge produces an *electrostatic force*. See **charge; static electricity**.

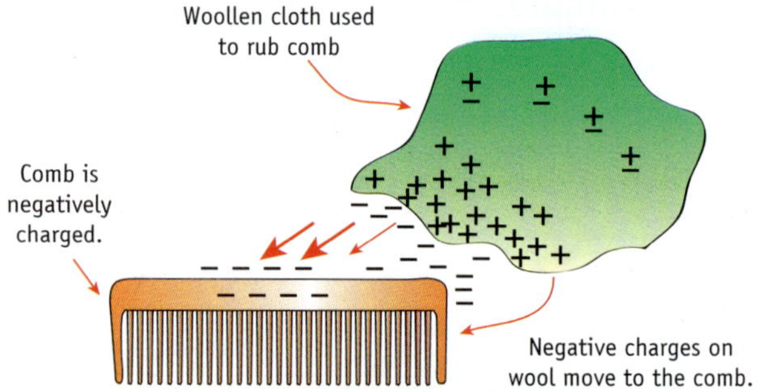

chemical energy

is stored in substances such as *food* and fuels. *Chemical reactions* release the *energy*. See **energy transformation**.

chemical reaction or change

occurs during chemical reactions, when particles in substances are rearranged to form a new substance(s) with different *properties*. For example, when wood burns, it combines with *oxygen* to form *carbon dioxide*. The stored *energy* in the wood is changed to *heat energy*, *light energy* and *sound energy*. A chemical change is often difficult to reverse. See **decomposition**.

chemical reaction, speed

can be fast (e.g. a match *burning*) or slow (a nail *rust*ing; milk going sour). It will go faster if the *temperature* is increased, the concentration of the chemicals is increased, the chemicals are ground into small pieces or a powder, or a *catalyst* is added.

Fast reactions

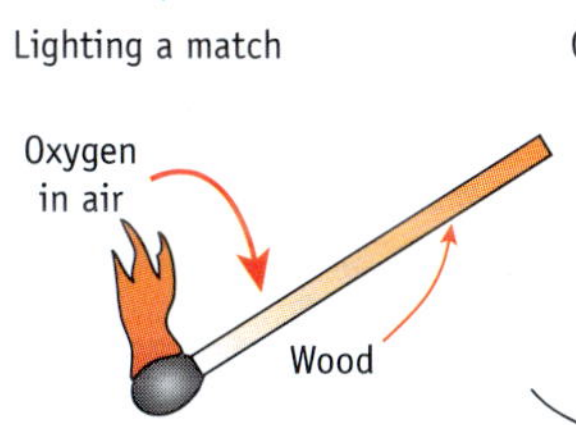

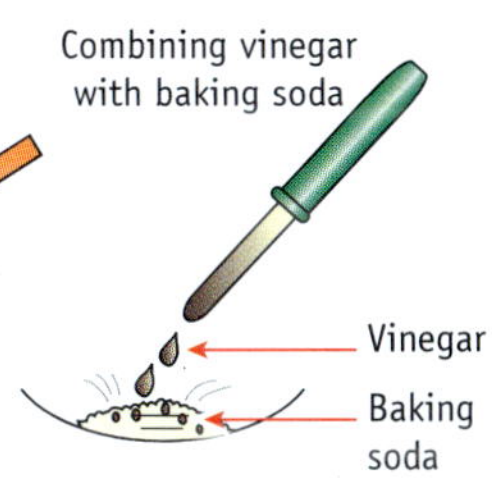

Slow reactions

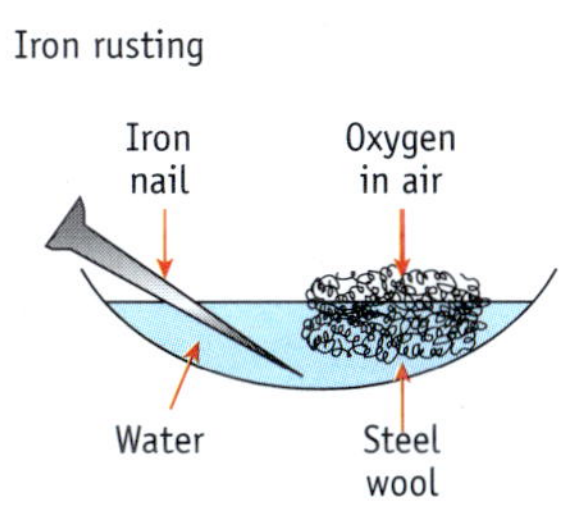

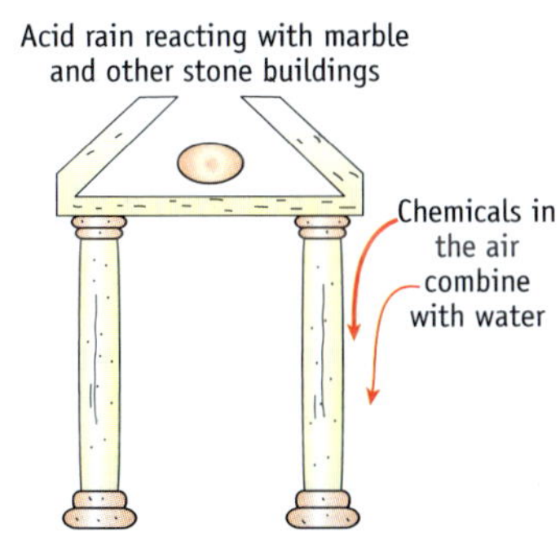

chemical symbols, elements

See **element symbols**.

chemical weathering

occurs when chemicals in water or the *air* react with *minerals* in *rock*. The chemical change weakens the rock so that it crumbles.

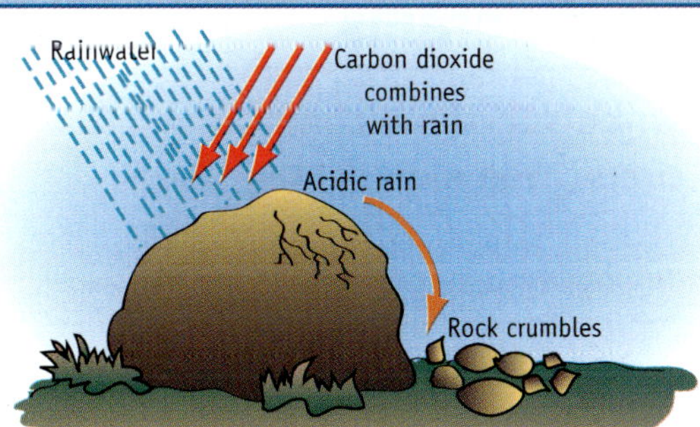

chloroplast

is a structure in plant *cells* where *photosynthesis* occurs. It contains green pigment that collects the *Sun*'s energy needed for this process. See **cells in living things**.

chromatography

is used to separate and detect small amounts of chemicals in *mixtures* (e.g. chemicals in urine samples). A drop of the mixture is placed on special absorbent paper, which is dipped into a *solvent*. The solvent then separates each chemical.

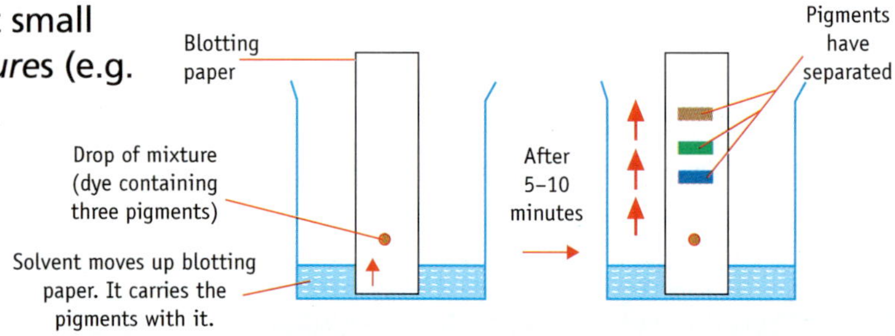

circuit

is the continuous pathway along which *electricity* flows. Electricity will not flow if there is a break or gap in the circuit. See **circuit diagrams; circuits in series; circuits in parallel, switch.**

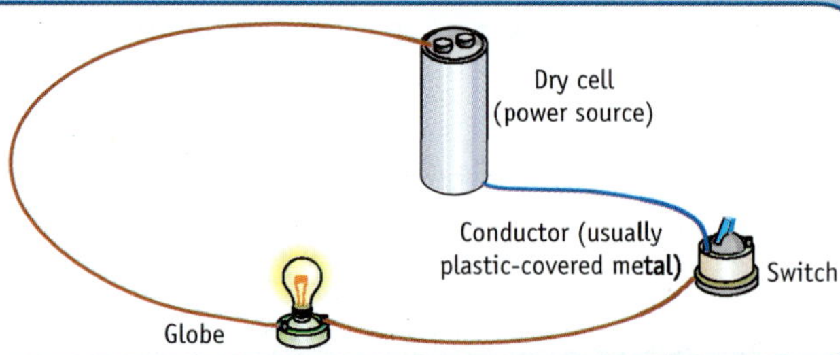

circuit, closed

See **switch**.

circuit, open

See **switch**.

circuit diagrams

use symbols to show components (parts) of a *circuit*. See **electrical symbols**.

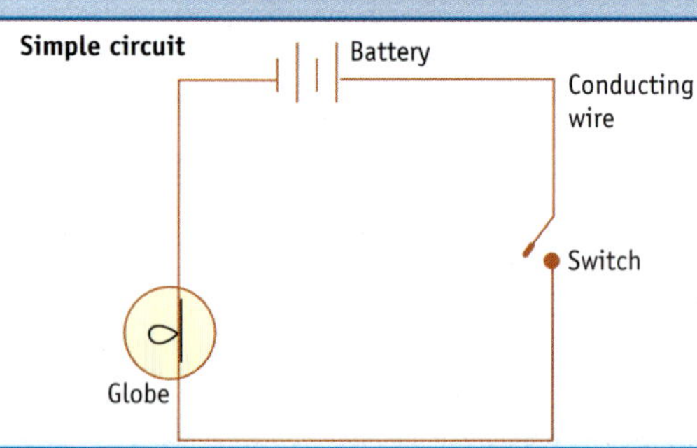

circuits in parallel

is when some of the components (or parts) in a *circuit* are parallel to each other. A meter to measure *voltage* is placed in parallel. Compare **circuits in series**. See **batteries in parallel**.

Globes in parallel glow brightly as the current passes through each globe separately.

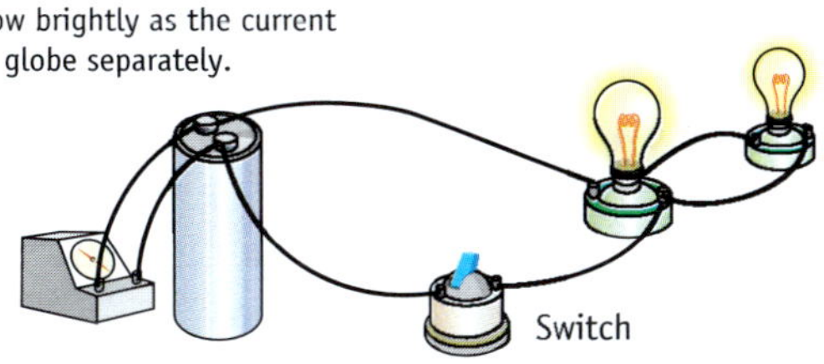

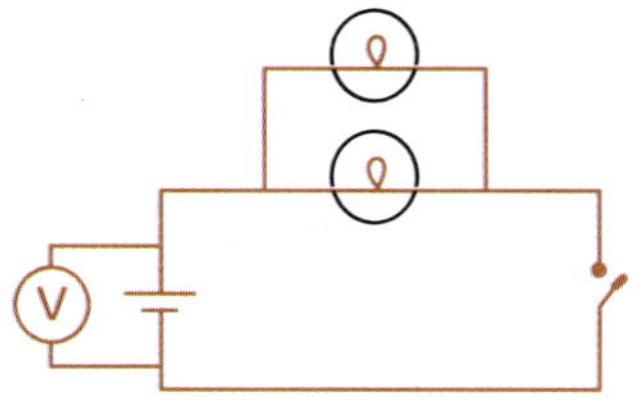

circuits in series

is when components (or parts) of the *circuit* are joined in line. Compare **circuits in parallel**. See **batteries in series**.

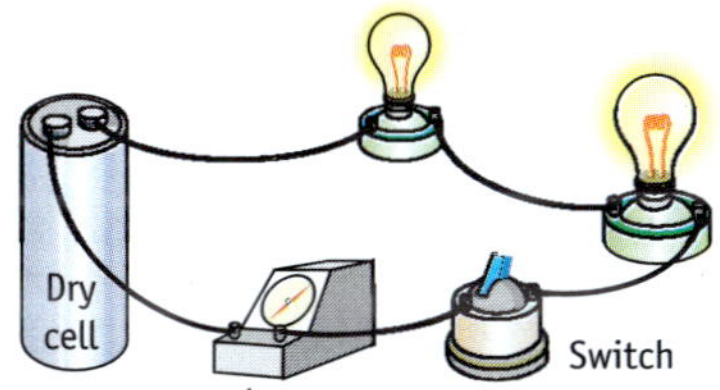

Globes in series glow dimly as the current is pushed through two globes.

Meter to measure current is always joined in series.

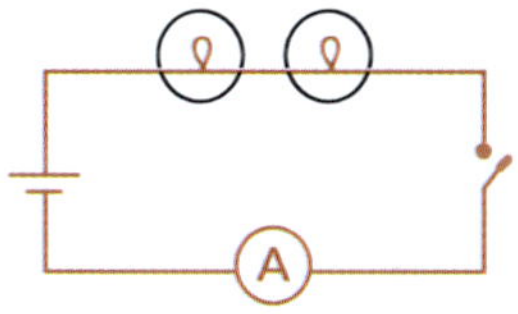

circulatory system

in a humans consists of heart, blood and vessels (arteries, veins and capillaries). The heart pumps blood around the body in blood vessels. The blood carries the *food* and *oxygen* to *cells* and collects wastes (e.g. *carbon dioxide*) and takes them to the *excretory system*.

Circulatory system

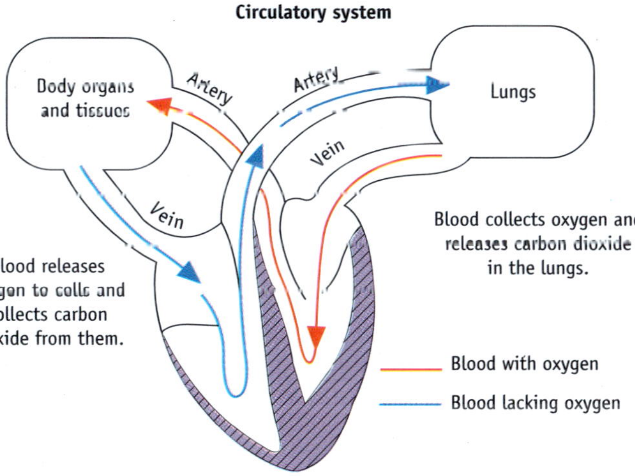

- Blood is carried away from the heart in arteries and to the heart in veins.
- Blood passes through the heart twice during one circulation of the body.

classification of living things

groups *living things* by their similarities in structure (e.g. body covering, *flower*s or cones produced, and number of legs). *Organism*s in the same group are usually closely related. See **kingdoms**; **species**; **keys**.

Jointed invertebrates have different numbers of legs.

Insects have six legs.

Spiders have eight legs.

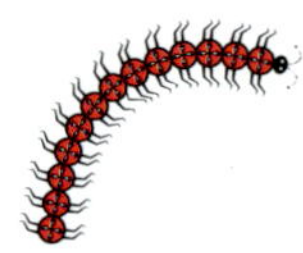

Millipedes have many legs.

Vertebrates have different body coverings.

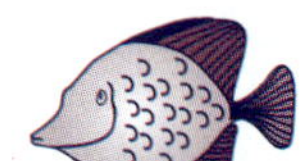

Fish have scales.

Birds have feathers.

Mammals have fur or hair.

climate

is the average *weather* conditions in an area over many *year*s. *Temperature* and *rain*fall are commonly used to define climate zones (e.g. tropical, dry or arid, temperate or mild, and snowy or polar). Climate influences the types of *plant*s and *animal*s that can live in an area.

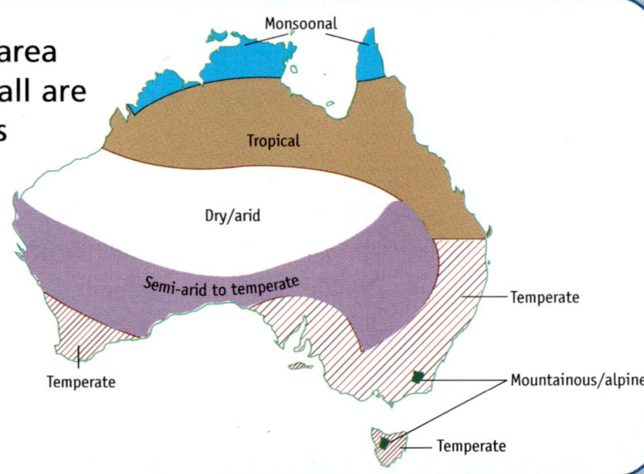

cloning

produces offspring that are identical to the parent. A breeder of *organism*s can select the features desired and then clone them (e.g. large, juicy corncobs).

cloud

is formed when moisture in the *air* cools and condenses on dust. Clouds are made of millions of very fine water droplets. See **cloud formation**; **cloud types**.

cloud formation

is the process of forming *clouds*. When warm *air* rises, it cools and moisture condenses to form *clouds*. Warm *air* rises by:

1. *convection*;
2. moving up a mountain;
3. meeting a mass of cold air.

See **cloud types**.

cloud types

are named by their height, shape and colour. Cirrus clouds are high, white, feathery strands. They are made up of ice. They usually indicate fine *weather*. Cumulus clouds are woolly and constantly change their shape. Small cumulus clouds indicate fine weather but if they grow large they may bring *rain* or thunder-storms. Stratus clouds are low, grey and uniform. They often produce drizzle.

Cirrus

Cumulus

Stratus

coal

is a *fossil fuel*. It is used for heating and to produce *electricity* and *gas*.

Plants fall into swamps where they cannot decay.

Sediments are compressed by layers of soil and so form peat.

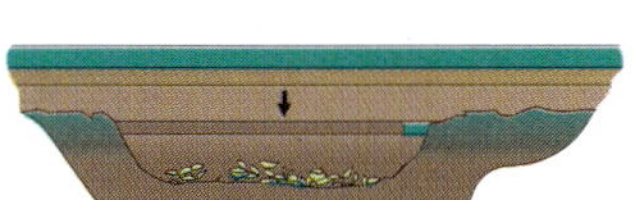
More compression changes the peat to coal.

coastal landforms

are made up of beaches joined to cliffs, with caves and rock platforms with rock pools.
See **erosion from water**.

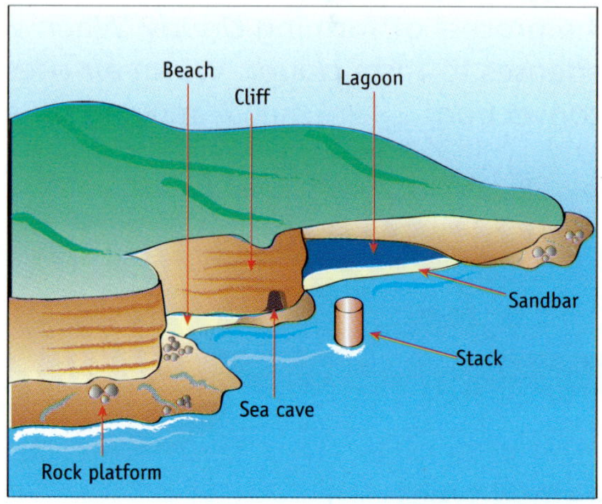

collaboration in ecosystems

See **cooperation or collaboration in ecosystems**.

colloid

is a *mixture* with the particles of substance finely and evenly spread through another substance (e.g. fog, paint). The particles will not settle out and cannot be separated by filtering. See **emulsion**.

colony

is a group of *organisms* in the same *species* living closely together. Termites, ants and some bees and wasps have complex societies. In termites, the reproductive male fertilises the queen (reproductive female). She produces *eggs* that hatch into workers; they build the nest, gather *food* and care for the young. Some become soldiers that protect the colony.

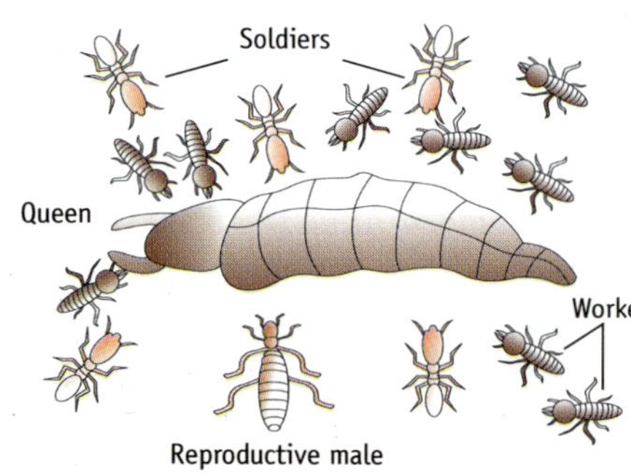

comet

is a body of dust, rock and frozen *gas* that travels around the *Sun*. When a comet is close to the Sun, it warms up and glows and forms a tail of dust and *gas*, which points away from the Sun.

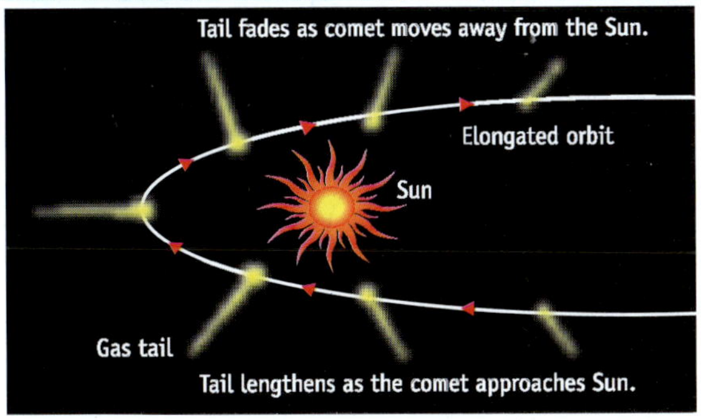

communication

is the transfer of ideas and information between *organisms*. Communication helps *animal*s defend themselves, and find *food* and mates. Animals use their *sense*s to communicate. See **courtship**; **communication between humans**.

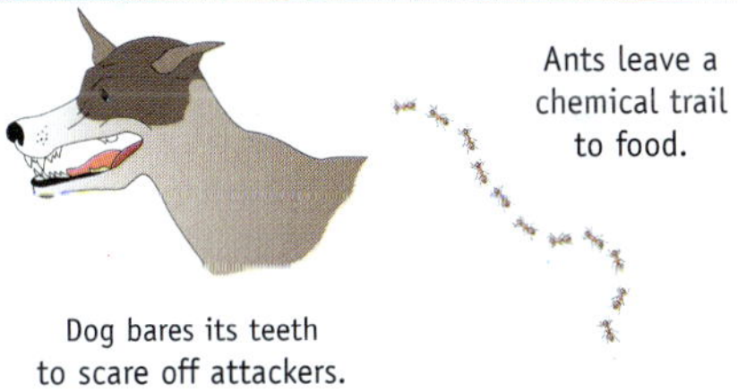

communication between humans

involves all the *sense*s as well as speech, body language and writing. Modern *technology* has given us new ways to communicate (e.g. newspapers, radios, telephones, *satellite*s and computers).

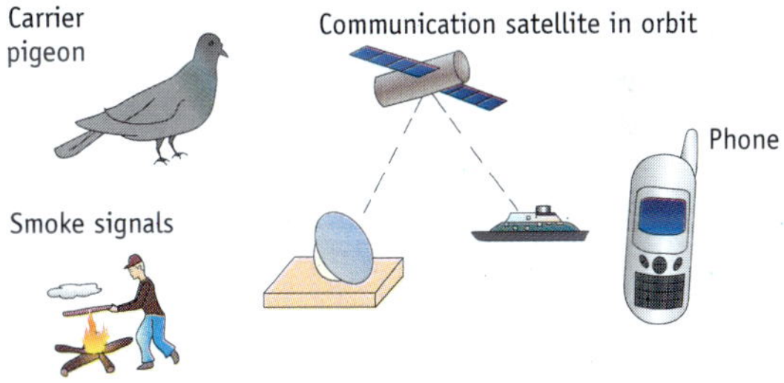

competition

occurs when *organism*s in an *ecosystem* fight each other for *food*, water, space and shelter, especially if they are in short supply or difficult to obtain.

components in parallel

See **circuits in parallel**.

components in series

See **circuits in series**.

compost

is a *mixture* of decaying *matter* such as leaves, grass and manure. It is useful for fertilising *soil*.

compound

is a chemical substance made of particles or molecules containing two or more *elements* joined by chemical *bonds*. Compounds can be broken down to simpler substances. Water, *carbon dioxide* and hydrogen *gas* are compounds. Compare **element; mixture**.

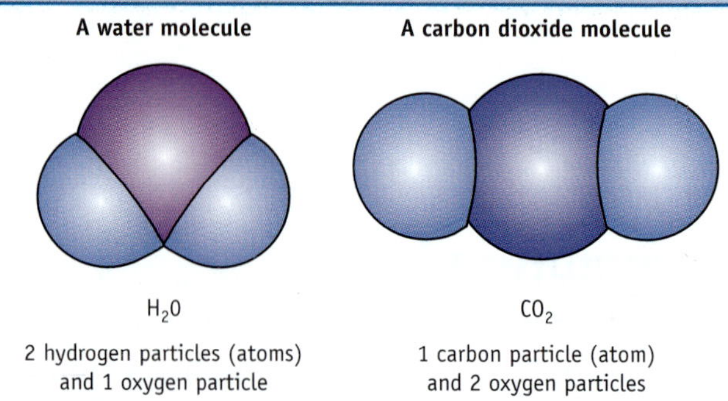

compression of matter

is when particles are pressed into a smaller space. *Gases* can be compressed, but *liquids* and *solids* cannot be compressed because their particles are already packed closely together.
See **particle theory of matter; pressure and the particle theory**.

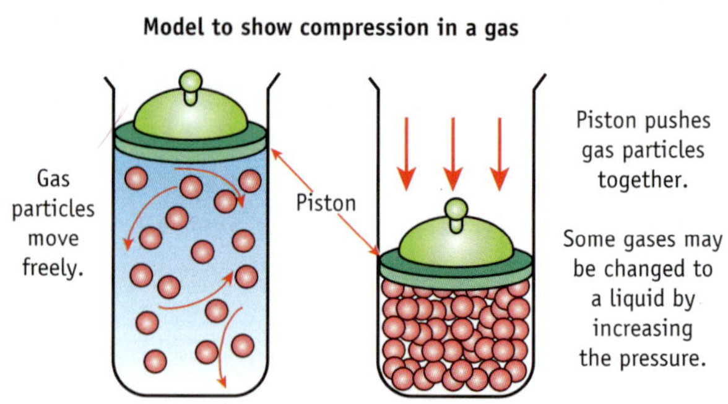

concentrated solution

has a large amount of *solid* or *solute* (e.g. sugar) *dissolved* in a *liquid* or *solvent* (e.g. water). Compare **dilute solution; saturated solution**.

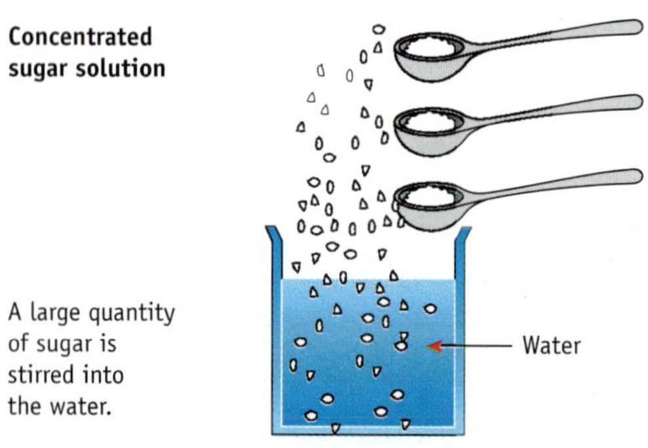

condensation

is a *change of state* from a *gas* to a *liquid* (e.g. water *vapour* on a cold surface will collect as water droplets). Condensation is used in *distillation*, for separating substances in *mixtures*.

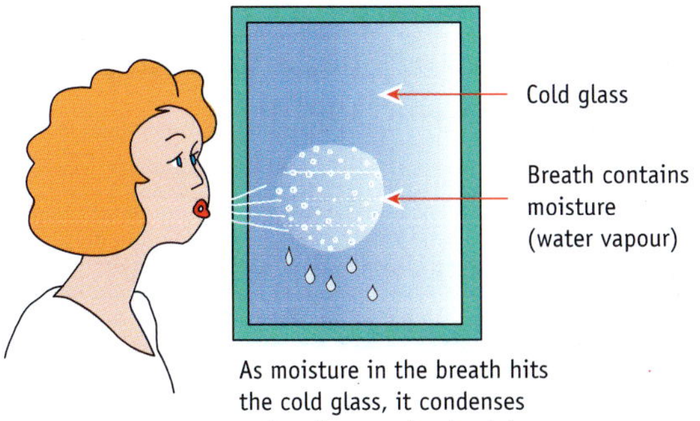

As moisture in the breath hits the cold glass, it condenses and so forms water droplets.

conduction of heat

is one way *heat* is moved (transferred) from one place to another. When a substance is heated, its particles vibrate more and pass on their *energy* by bumping the next particles. See **conductors of heat**.

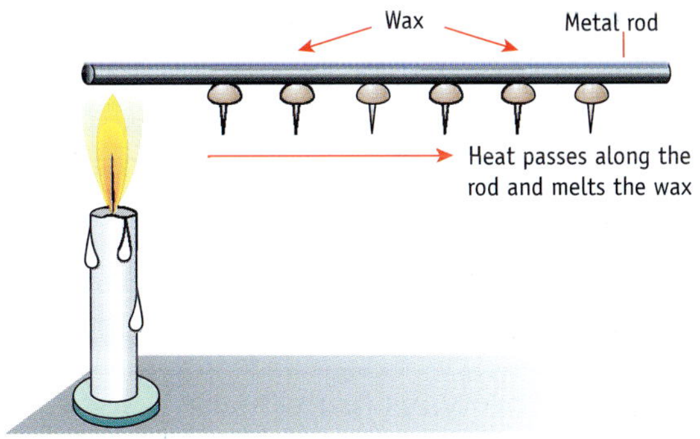

conductors of electricity

transfer *electricity* from one place to another. All *metals* conduct electricity. Copper is commonly used to carry *currents* in *circuits*. See **insulator (electricity)**.

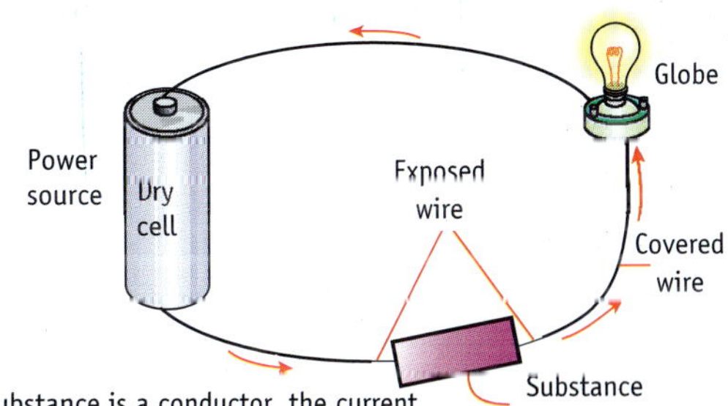

If the substance is a conductor, the current will flow through it and the globe will glow.

conductors of heat

transfer *heat* from one place to another. *Metals* are good conductors of heat, but most other *solids*, *liquids* and *gases* are generally poor conductors. See **conduction of heat; insulator (heat)**.

conglomerate (puddingstone)

is a *sedimentary rock* containing pebbles, which have been deposited by fast flowing water.

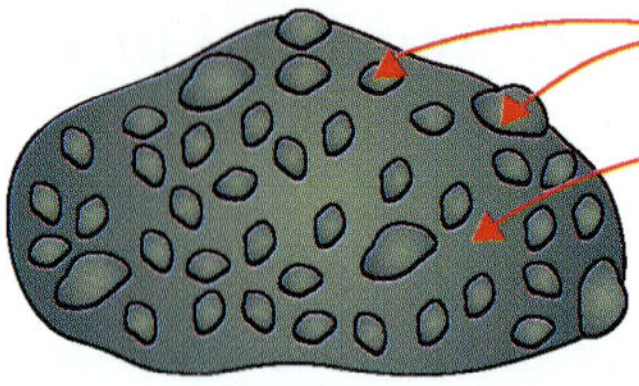

conservation

means preserving or protecting something (e.g. *energy* or the *environment*).

conservation of energy

means the total amount of *energy* in the *Universe* remains the same. Energy is not created or destroyed. It may be moved to another place or changed into another form of energy. See **energy transfer; energy transformation; conservation of non-renewable resources**.

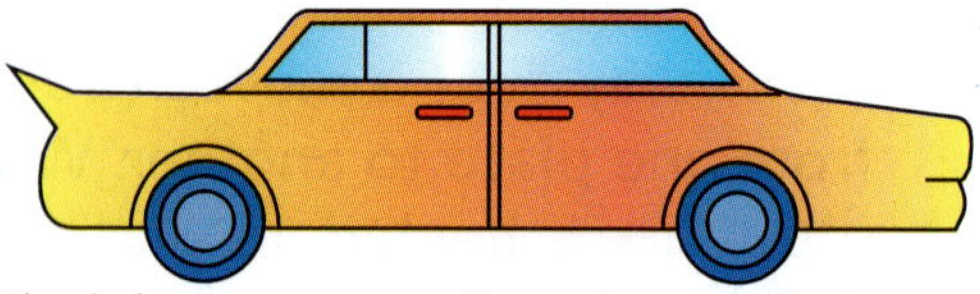

conservation of the environment

refers to the ways humans care for and protect the land, the *environment* and its *resources*. Many industries are now managed so that they help with conservation. For example, the timber industry regenerates forests and restricts wood harvesting. See **sustainable**.

conservation of fossil fuels

See **conservation of non-renewable resources; conservation of petrol**.

conservation of mass

During *physical* and chemical changes, no *matter* (*mass*) is created, lost or destroyed. The total mass remains unchanged.

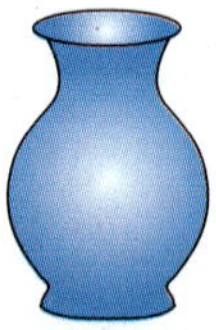

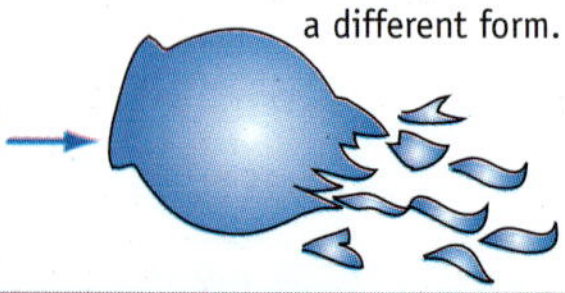

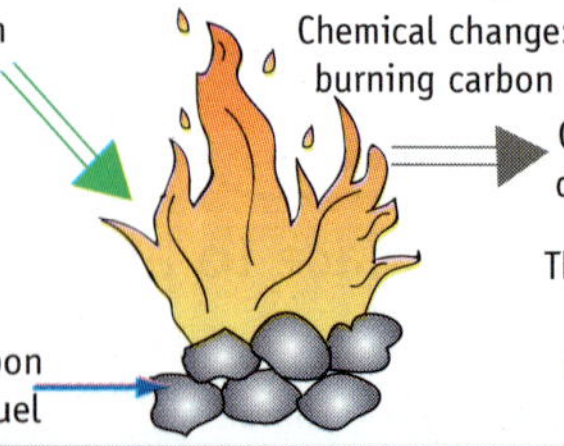

conservation of non-renewable resources

means conserving (protecting) *materials* by recycling and reusing them, or using materials that are renewable (e.g. using natural fibres like wool and cotton instead of *synthetics*, which are made from *fossil fuel*). See **sustainable; conservation of petrol; non-renewable resources**.

conservation of petrol

is when alternatives are used to replace petrol in cars. They include *electricity* from batteries, alcohol produced from crops, *gas* from *animal* manure or rotting garbage, and oils from *seeds* and *fruits* or oil *shale*. See **sustainable**.

Using muscles conserves petrol.

conservation of soil

See **soil conservation**.

constellation

is a cluster of *stars* forming a pattern or shape. Ancient people saw these patterns as pictures (e.g. scorpion, bear, Orion the hunter, the Southern Cross). Modern star charts show 88 constellations in the sky.

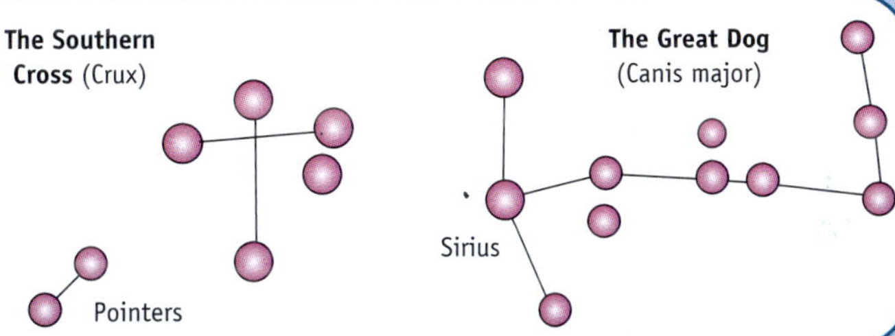

consumers

are *living things* that obtain their *energy* by feeding on other living things. *Animals* are consumers and they may be *herbivores*, *carnivores* or *omnivores*. See **food chain**. Compare **producers; decomposers**.

contraction of matter

occurs when *matter* is cooled. It takes up less space. Cooling slows down the movement of particles so that they do not bounce off each other as much and they spread out less. See **particle theory of matter**.

Power lines sag in hot weather.
In winter they contract and shorten.

control (in an experiment or fair test)

is used as a comparison. In an *experiment* to see if water evaporates faster in *wind*, a wet cloth is set up in wind and another (the control) is set up without wind. They are then compared. See **scientific method**.

convection

is one way *heat* travels in *liquid*s and *gas*es. When a *gas* or liquid is heated, the particles move faster and spread out. They become less dense and rise. Cooler, denser particles move in to take their place. Compare **conduction; radiation of heat**. See **convection in the atmosphere**.

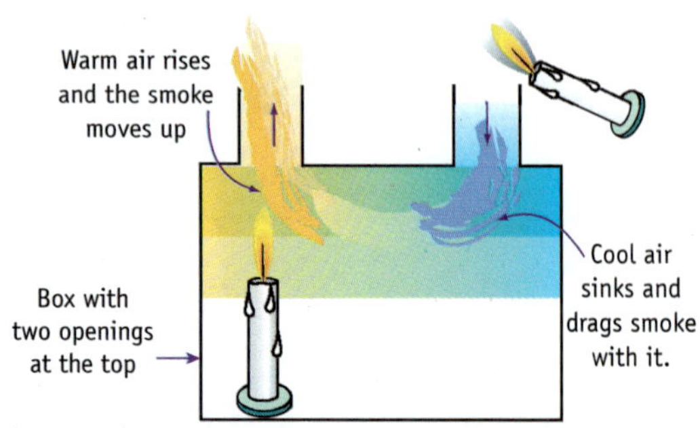

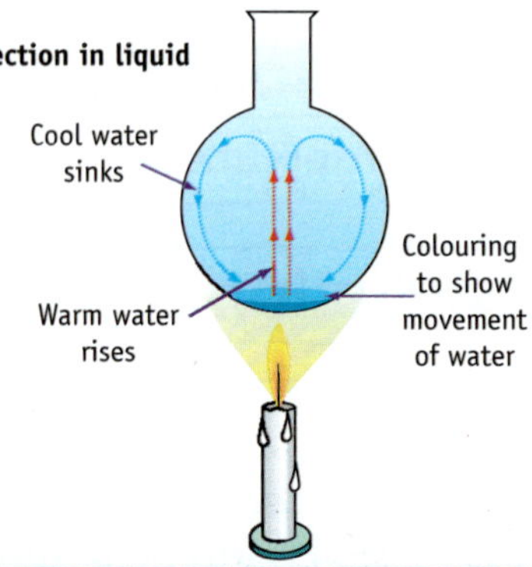

convection in the atmosphere

occurs when the *Sun* heats the *air* near the ground. This makes the air expand and become less dense (lighter), so that it rises. The *temperature* of the atmosphere decreases with height from the *Earth*. As the air rises, it cools, becomes heavier and then falls. See **global air currents; clouds**.

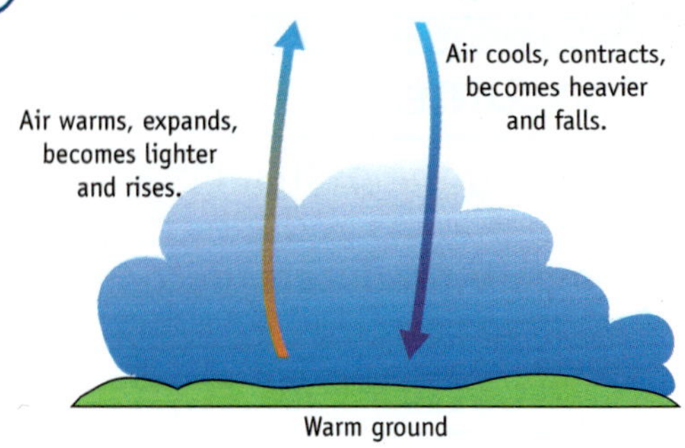

cooperation or collaboration in ecosystems

is when *organism*s work together to increase their survival rate (e.g. wolves hunt in packs; male yaks surround the females and young when *predators* threaten; dolphins surround schools of fish to catch them). See **colony**.

core

is at the central region of a planet. See **Earth**.

corrasion

occurs when *rock* fragments carried by *wind*, water or ice beat against the rocks or grind into them, causing them to crumble. See **erosion from wind; erosion from water**.

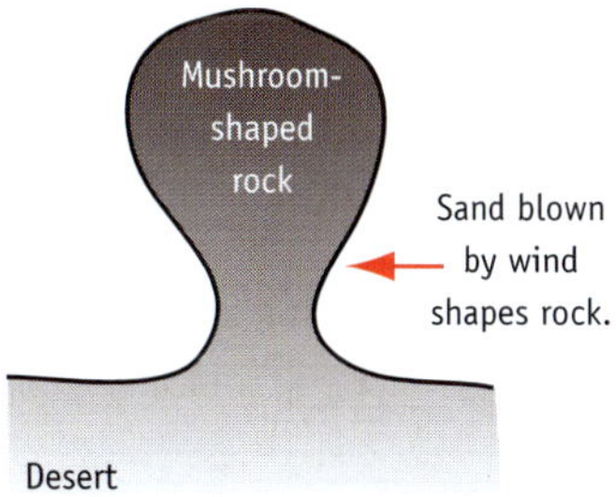

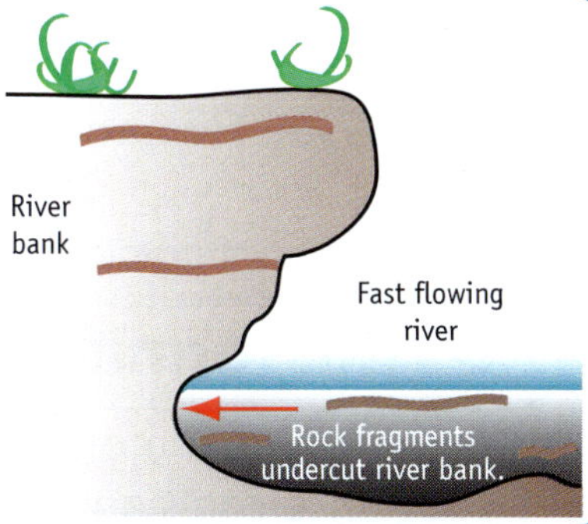

corrosion

occurs when a *metal* reacts with other chemicals. For example, iron slowly combines with *oxygen* and water to form *rust*.

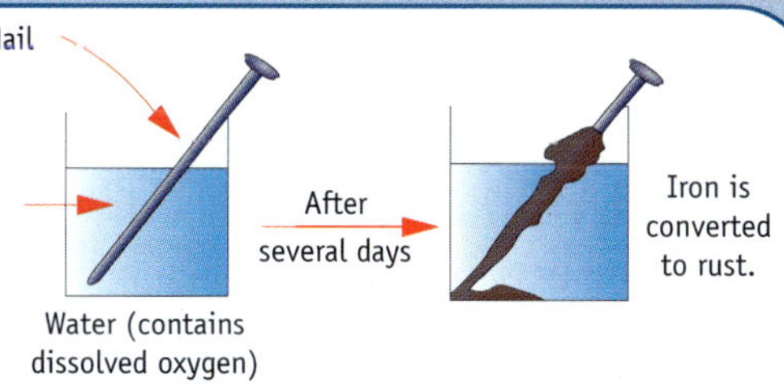

Word equation:

iron + oxygen + water ⟶ hydrated iron oxide (rust)
See **rust proofing**.

courtship

is behaviour that occurs before mating. It ensures *animals* mate with the same *species* and encourages the female to receive the male. In birds, courtship often includes elaborate song, display of feathers and dancing. For example, the male lyre bird mimics sounds and performs elaborate displays.

The bower bird builds a bower for mating. He decorates it with coloured objects such as feathers, berries, leaves and shells.

crust

is the outermost solid layer of the *Earth* and is made up of the oceanic crust and continental crust.

crystal

is a *solid* substance that contains regular shapes with flat surfaces and sharp edges. Crystals may be microscopic. The particles in a crystal are arranged in a lattice. See **crystal formation**.

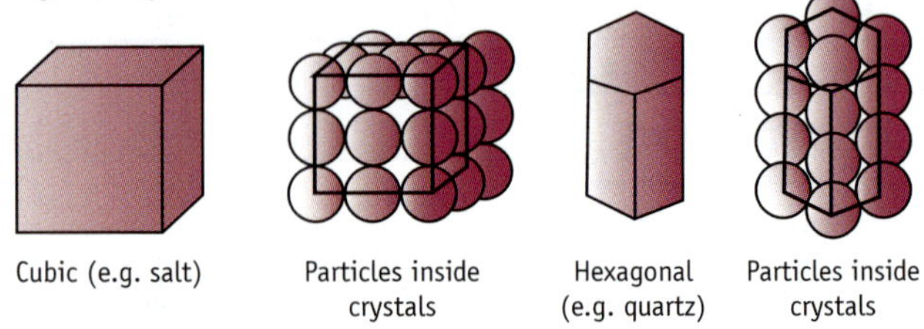

The arrangement of particles inside the crystal gives it shape.

crystal formation (crystallisation)

occurs when a molten substance cools and solidifies, and when a *liquid* is evaporated from a *solution* (e.g. salt water). The slower the process, the larger the *crystal*s. See **igneous rocks; crystal**.

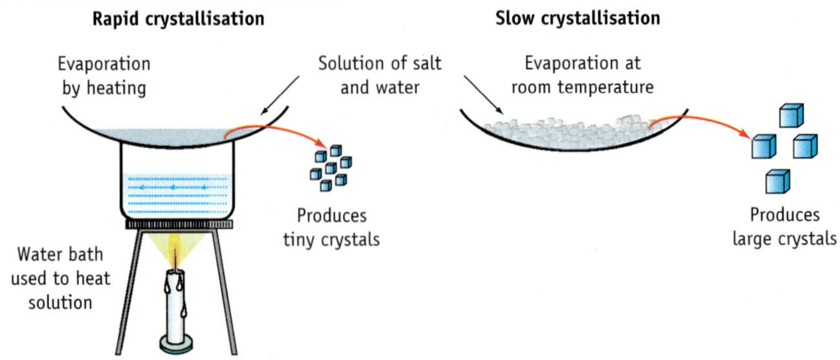

crystallisation

See **crystal formation**.

current

See **electric current**; **current electricity**; **global air current**; **ocean currents.**

current electricity

is moving *charge*. Electricity is produced by batteries and *generator*s and is transferred from one place to another by a *current* in an electric *circuit*. Compare **static electricity**.

cyclone

is a very intense whirlpool of *wind* that forms over the ocean. It can travel at *speed*s greater than 200 km/hr and can also cause huge waves and massive damage.

dams

should be built on *rock* where water will not leak away. Surrounding and base rock must provide a suitable foundation to support the wall. The wall needs to be strong to hold back the water. See **foundations**.

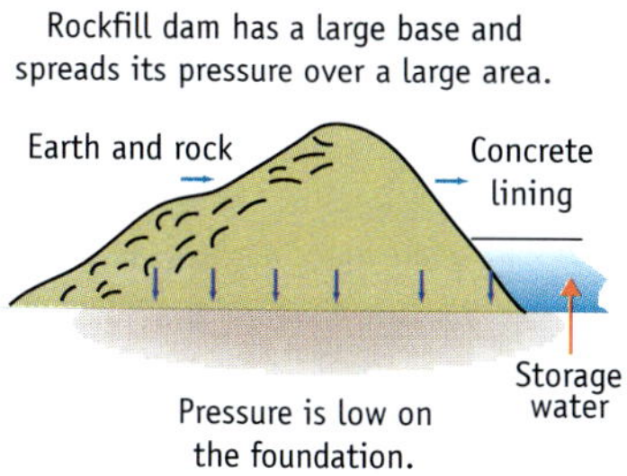

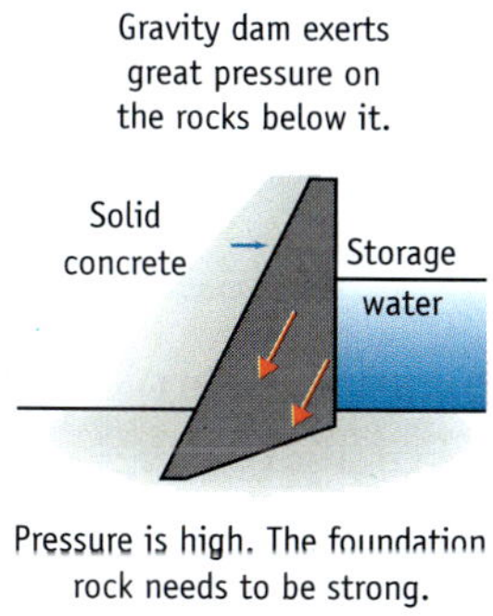

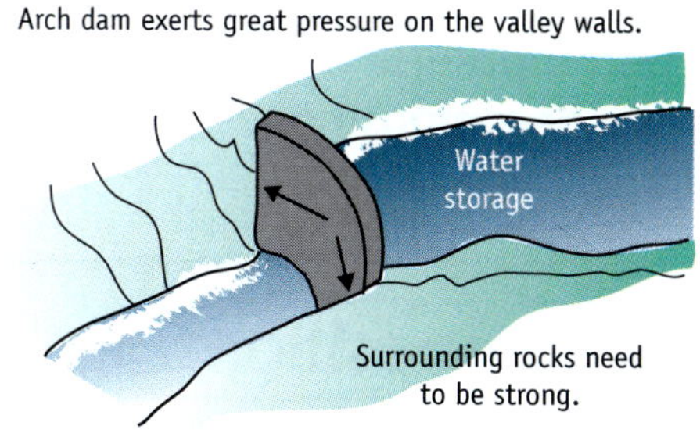

day and night

is the length of time for a planet to spin or rotate on its axis. The side facing the *Sun* is in daytime. The side away from the Sun is in darkness or night. The Earth rotates once in about twenty-four hours.

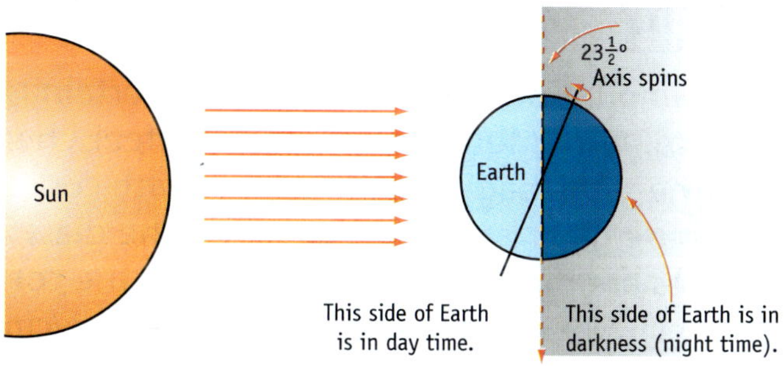

decanting

See **sedimentation and decanting**.

deciduous

means a tree or shrub sheds its leaves, usually during autumn. Compare **evergreen**.

decomposers

break down dead *plants* or *animals* and their wastes and cause them to decay, releasing *nutrients* into the *soil*. *Bacteria* and *fungi* are decomposers.
See **decomposition**.

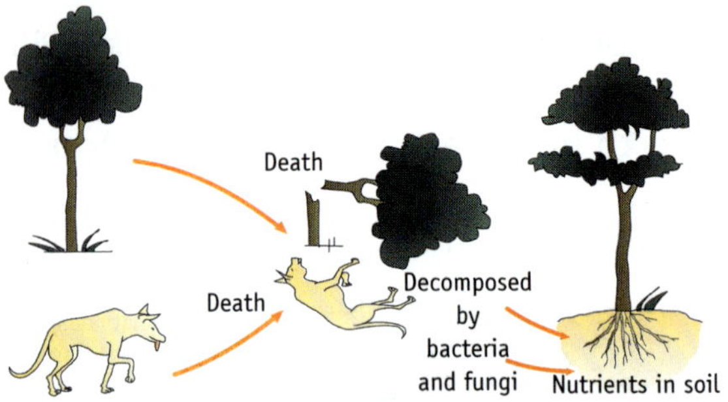

decomposition

is a *chemical reaction* where a substance is broken into two or more new substances. For instance, heating sugar breaks it down to carbon and water. See **decomposers**.

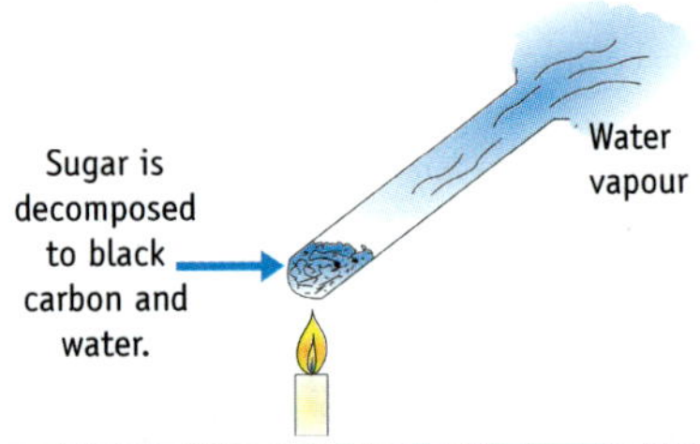

density

is the amount of *matter* or *mass* in a certain *volume* or space. A bag full of glass marbles 'feels' heavier than the same bag when it is filled with corks. This is because glass is denser than cork (the glass particles are more closely packed than the cork particles). Density is measured in g/mL or g/cm^3.

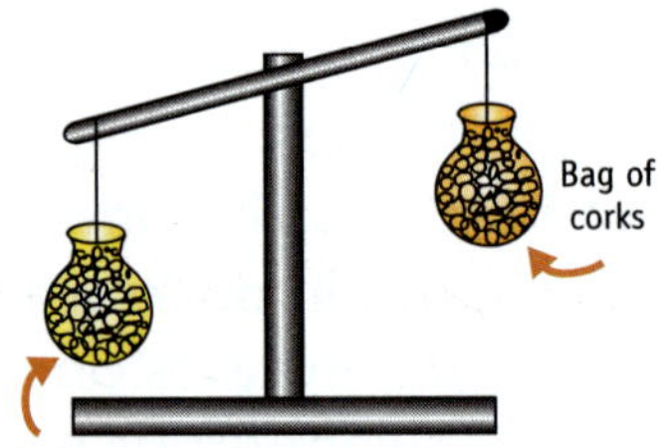

Bag of marbles has more mass than a similar bag of corks because the marbles are more dense.

desert landforms

are made up of plains with smooth polished pebbles (gibbers), hard rock hills (mesas and inselbergs), braided (branched) streams and salt lakes.

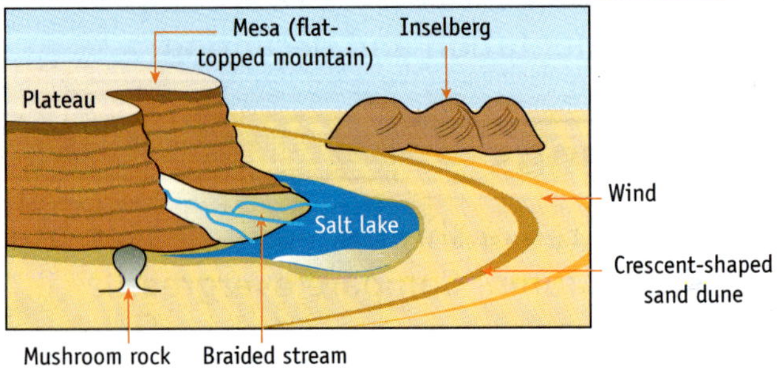

detergents

See **soaps and detergents**.

diffusion

is the spreading and mixing of one substance into another. Diffusion is caused by the natural movement of particles.
Diffusion *speeds* up if the *temperature* increases.

Cooking smells spread by diffusion.

Smells of perfume spread by diffusion.

Dissolving tablet spreads slowly through the water by diffusion.

digestion

(in humans) occurs when *food* is broken down in the *digestive system* into small particles so that it can pass through the intestine walls into the blood. See **digestive system; digestion, mechanical; digestion, chemical**.

digestion, chemical

occurs in the mouth, stomach and small intestines. Chemicals called *enzymes* react with *food* and break it into small particles. In the stomach hydrochloric *acid* helps with this digestion. See **digestion, mechanical; digestive system**.

digestion, mechanical

is the breaking up of *food* into small pieces. The teeth grind food into small pieces. Bile breaks *fat*s into small lumps. See **digestion, chemical; digestive system**.

digestive system

(in humans) breaks down *food* into small particles so that they can pass through the walls of the small intestines, into the blood. See **digestion**.

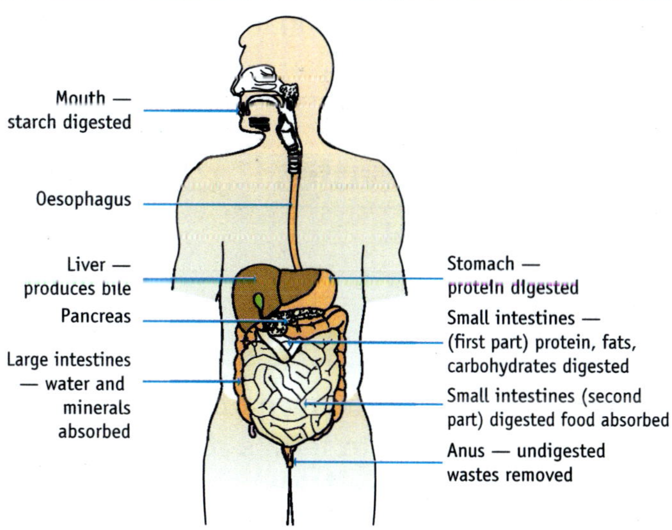

dilute solution

has a small amount of *solid dissolve*d in the *liquid*. Compare **concentrated solution; saturated solution**.

Dilute sugar solution

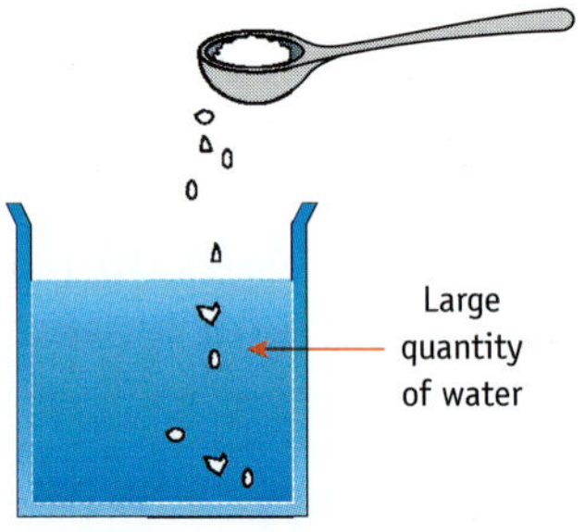

dinosaurs

were prehistoric reptiles that lived about 150 million years ago. *Fossil* remains show that dinosaurs ranged from the size of a chicken to 30 metres in length.

direct current

is a steady *current* flowing in one direction. It is produced by batteries. Compare **alternating current (AC)**.

disposal of wastes

See **waste disposal**.

dissolve

is to break down into tiny particles that are too small to see, as when a *solid* dissolves in a *liquid* to form a *solution*. For example, sugar stirred in hot water dissolves. See **soluble**.

distillation

is a method of separating *liquid*s from *mixture*s. The *solution* is heated and as each substance boils at a different temperature, the *vapour*s formed can be condensed and collected separately. Distillation is used to obtain fresh water from sea water, to separate perfumes and oils from *plant materials*, to collect petrol, kerosene and diesel oil from *petroleum* and to purify and concentrate alcohols.

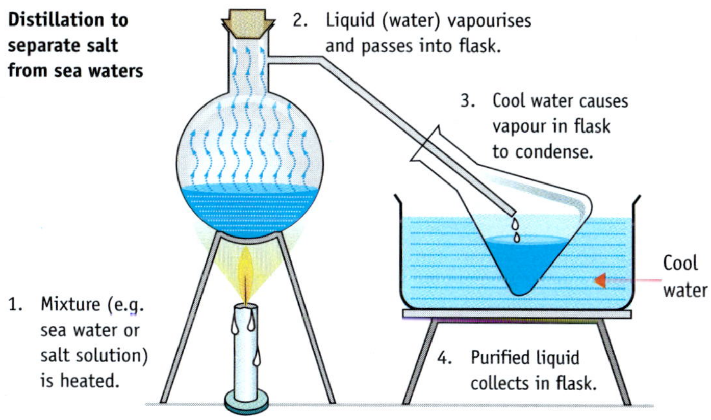

diurnal

means active during the *day*. Most animals are diurnal. Compare **nocturnal**.

diversity

See **biodiversity**.

DNA

is a chemical found in the nucleus of the *cell* in structures called genes. It controls the way animals develop and their features. Genes can be passed on from one generation to the next.

These people appear different because they have different DNA.

drought

occurs when there is a long period without *rain*. Droughts have a devastating effect on the natural *environment* and on farming and the production of *food*.

Areas frequently affected by drought

dry cell

See **electric cell**.

Earth

is the third *planet* from the *Sun*. It is a sphere slightly flattened at the poles and bulging at the equator. The outer zones of the Earth are the *atmosphere*, *lithosphere*, *hydrosphere* and *asthenosphere*.

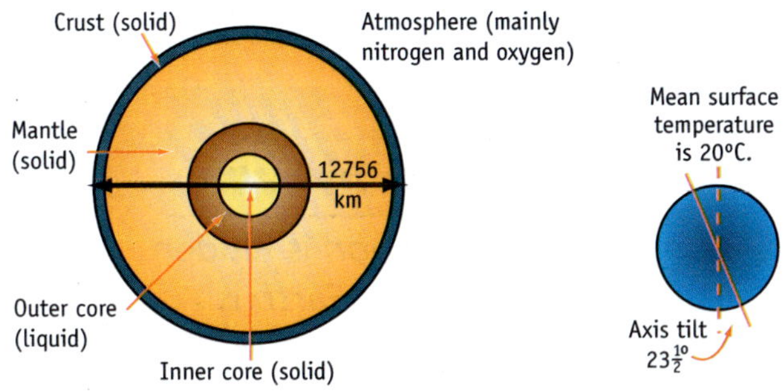

earthquake zones

are areas in which *earthquakes* occur. Most volcanic activity also occurs in these zones.
See **volcano**.

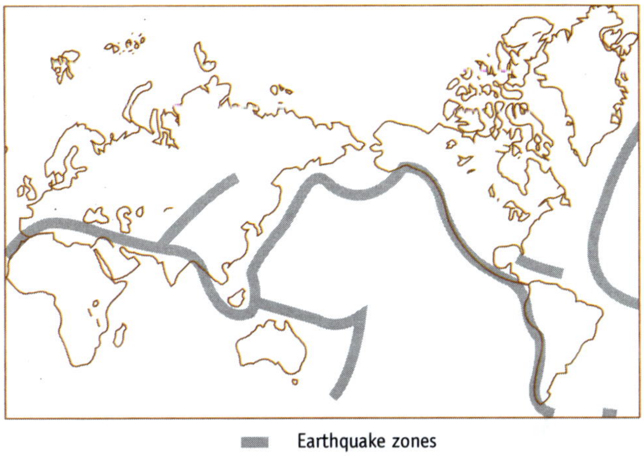

earthquakes

are vibrations in the Earth's *crust* caused by *Earth* movements. They are measured with *seismographs*. See **earthquake zones**.

echo

occurs when *sound* bounces off hard objects and is heard a split second later. Bats and dolphins use sound waves and echoes to find their way around. Ships use echoes to find the depth of the water.

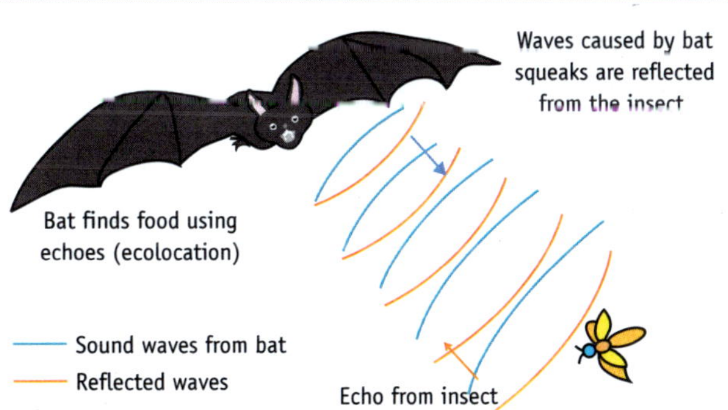

E

eclipse

occurs when one body in the sky hides another. See **eclipse of the Sun; eclipse of the Moon**.

eclipse of the Moon

occurs when the *Moon* is hidden. This occurs when the *Earth* is between the *Sun* and the Moon so that the Earth is blocking the Sun's rays and the Moon does not shine.

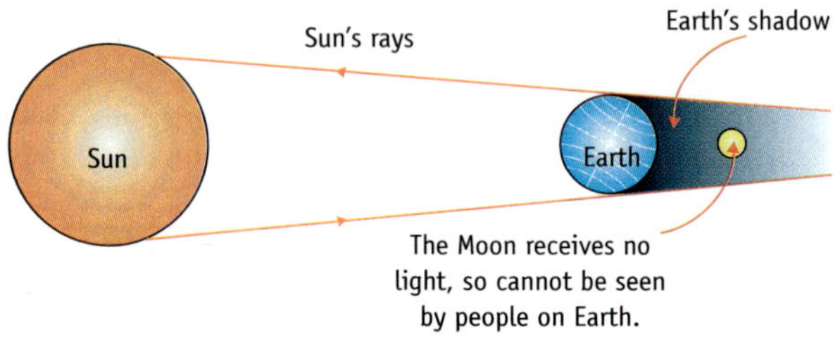

eclipse of the Sun

occurs when the *Moon* hides the *Sun*. This occurs when the Moon passes between the Sun and the *Earth*.

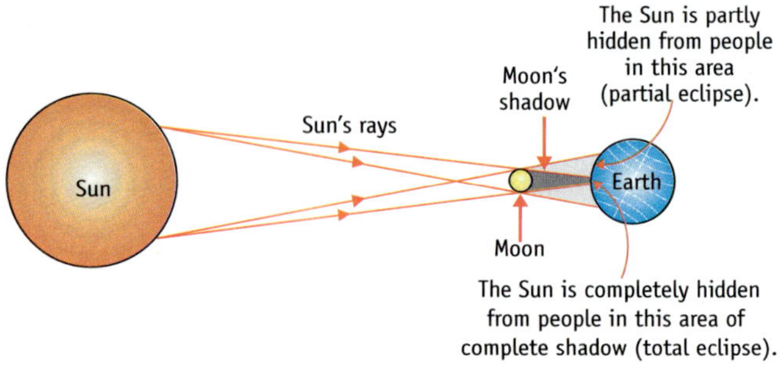

ecosystem

is a place containing *living things* that interact with each other and their surroundings (*environment*). It is made up of living and non-living features. A pond is an example of an ecosystem. See **ecosystem interactions**.

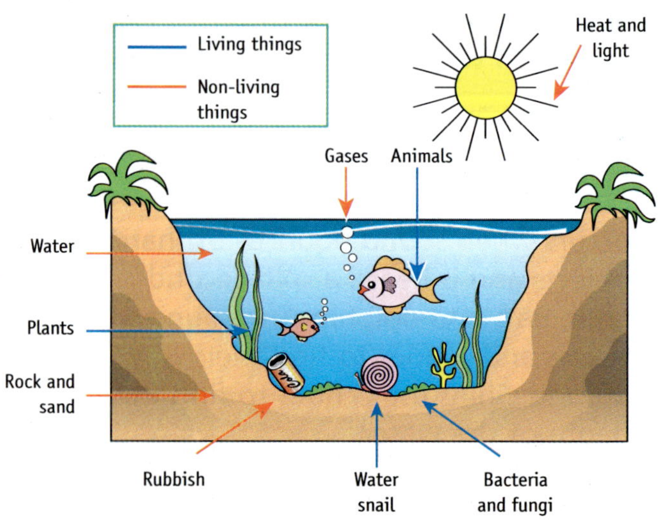

ecosystem interactions

occur when *organisms* in an *ecosystem* affect each other.

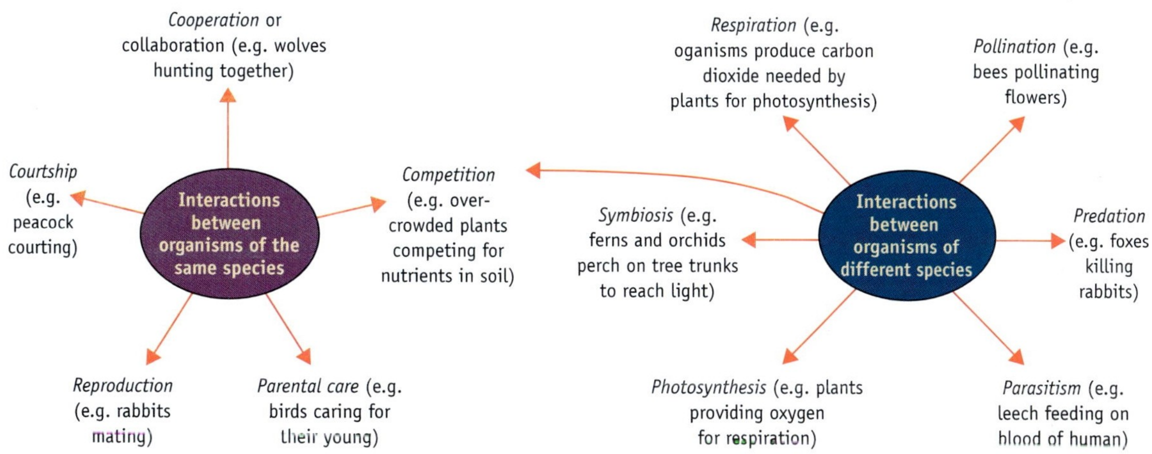

efficiency of a machine

is a measure of how well it uses *energy*. The energy (or *work*) we get from a *machine* is generally less than the energy (or work) we put into it. This is because *friction* changes some of the original energy to *heat energy*.

Some of the energy is converted to heat energy because of friction.

$$\text{Efficiency of a machine} = \frac{\text{work done by the machine}}{\text{work put into the machine}} \times 100$$

See **mechanical advantage**.

egg (ovum)

is the female reproductive or sex *cell*. In some *animals* it is enclosed in a shell with *food* reserves after *fertilisation*. It is produced by the ovary. See **fertilisation; reproductive adaptations**.

E

El Niño

is a weather pattern that occurs every three to seven years when the *wind* and *ocean currents* in the Pacific Ocean change. This alters the *climate* and causes *natural disasters*, such as *droughts*, floods and severe storms.

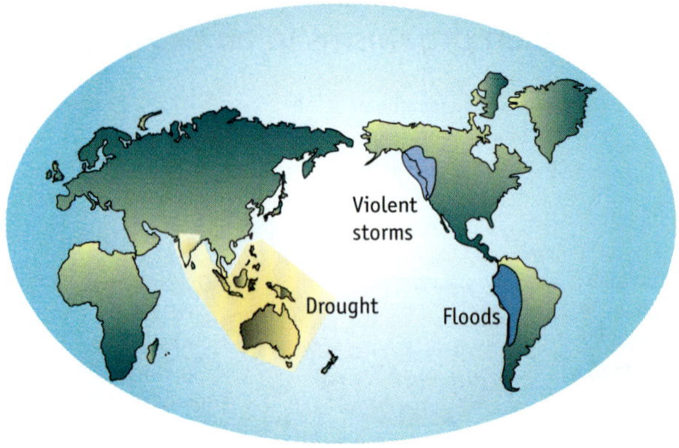

elastic material

is material that will return to its original shape after it has been stretched, twisted or bent (e.g. rubber). The material possesses elasticity.

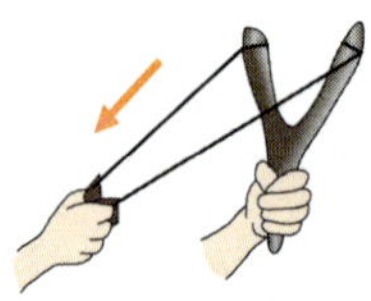

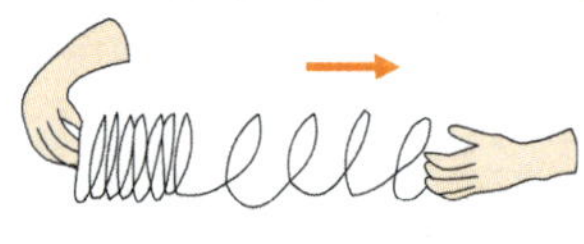

Rubber and springs have elasticity.

elasticity

See **elastic material**.

electric cells

are used to make up a *battery*. *Chemical reactions* produce a positive *charge* on the carbon rod and a negative charge on the zinc casing.

Inside a dry cell

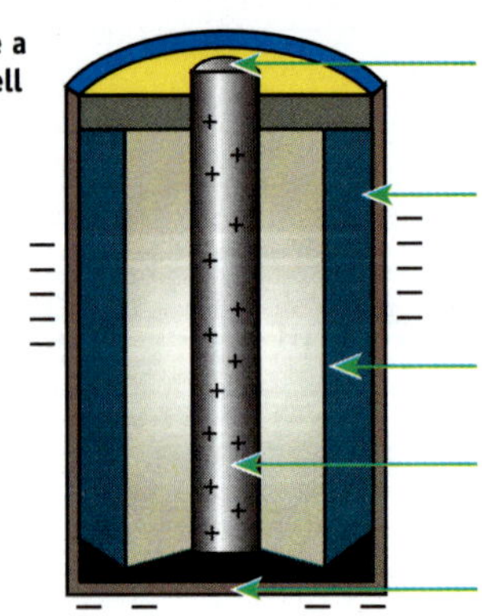

electric circuit

See **circuit**.

electric current

is a flow of *electricity* or moving negative *charge*. Electricity will move along a wire if there is a difference in electrical pressure or *voltage*. Current can be increased by increasing the voltage.
See **direct current (DC); alternating current (AC)**.

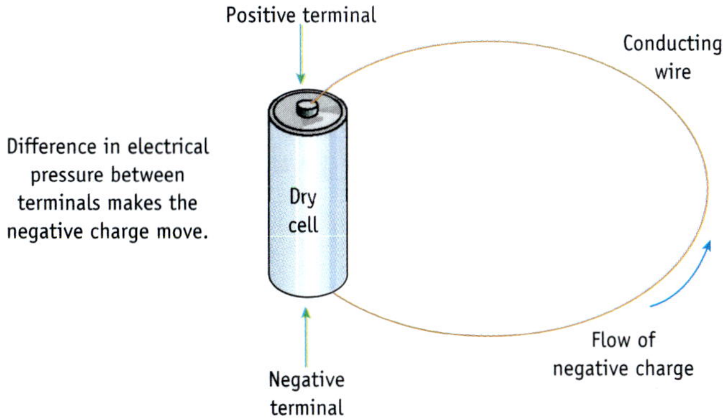

electrical energy

See **electricity**.

electrical energy used by an appliance

is shown by the *watts* used. The watts show how much *power* it draws from the *electricity* supply. The amount of *energy* used (in kilowatt hours) = electricity it draws (in kilowatts) x time it is in use (in hours).

What does a 3000 watt radiator cost to run for 2 hours?

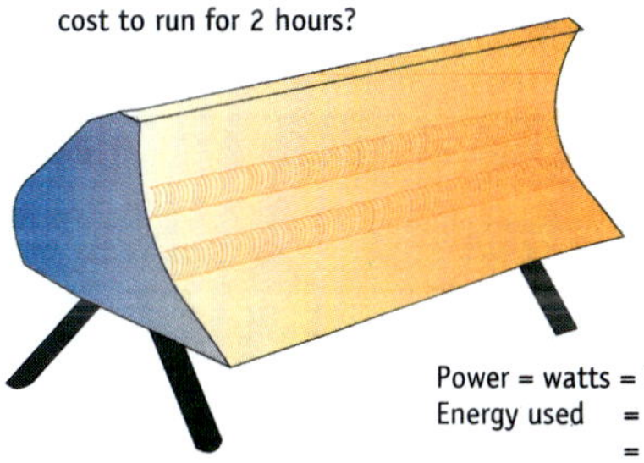

Power = watts = 3 kilowatts
Energy used = Power × time (in use)
= 3×2
= 6 kilowatt hours (kWh)
If 1 kWh of energy costs 10 cents, then 6 kWh will cost 60 cents.

electrical symbols

are used to show components (parts) of a *circuit*.

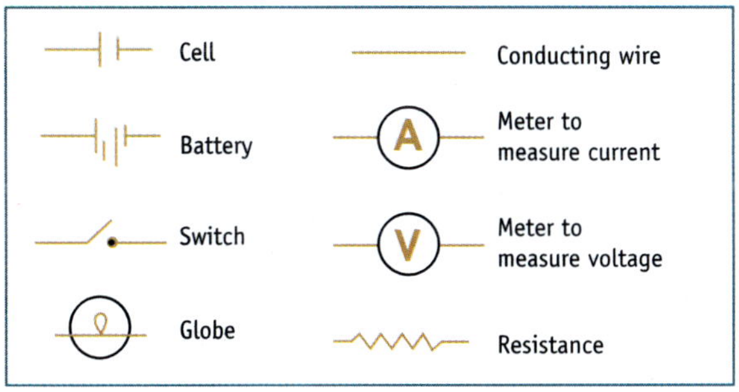

electricity

is a form of energy, as it can bring about a change in things. See **current electricity; static electricity; electrical energy used by an appliance**.

Electricity makes a globe filament heat and glow.

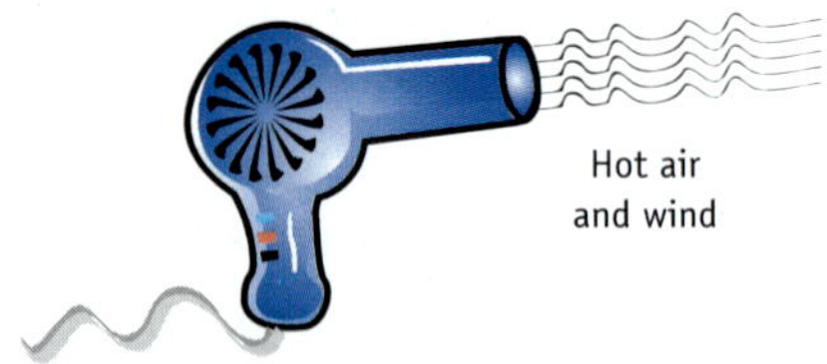

Electricity makes a hair dryer heat up and blow hot air.

electromagnet

is a *magnet* that can be switched on and off. Electromagnets are made by passing a *current* through a coil of wire wrapped around a piece of iron. Electromagnets are used in motors, speakers and electric bells, and on cranes to pick up and transfer large pieces of steel and scrap iron.

A simple electromagnet

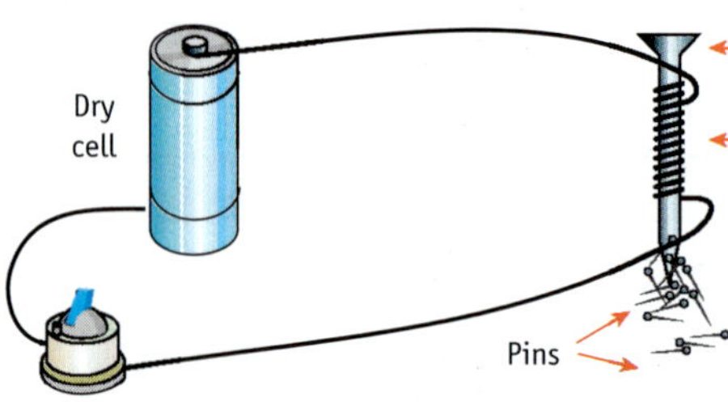

When current is on, the nail becomes a magnet. When the current is off, the pins drop.

electrostatic charge

See **charge; charging an object**.

electrostatic (electric) field

is the area around a charged object where it exerts a force. See **charge; static electricity; electrostatic force**.

electrostatic force

is the pull or push exerted by a charged object. For example, nylon clothes become charged in a dryer and cling to other clothes. Unlike charges attract each other. Like charges repel each other. See **static electricity; charge**.

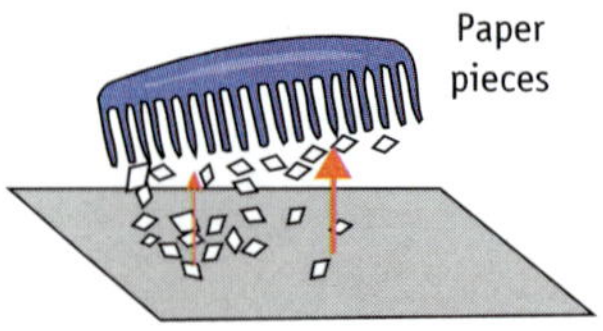

Comb and paper have opposite charges. The paper is attracted to the comb.

Each hair has the same charge.

element

is a pure substance made up of one type of particle or *atom* (e.g. the element carbon is composed only of carbon particles). It cannot be converted to a simpler substance. There are 92 naturally occurring elements. Scientists have made more by artificial means (e.g. plutonium and einsteinium). Most elements exist as solids but some are liquids or gases. See **element symbols**.

element symbols

are given to *element*s (each has its own symbol). Elements exist as *metal*s or *non-metals*. A small number of elements have *properties* between metals and non-metals (e.g. silicon, Si).

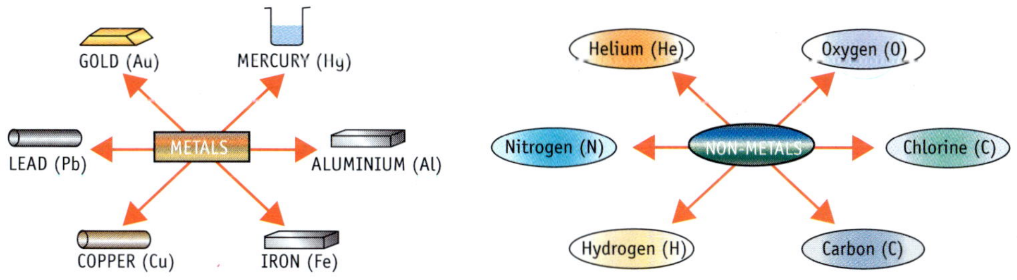

emulsion

is a *colloid* in which a *liquid* is spread through another liquid. For example, homogenised milk has tiny globules of butterfat spread through it.

energy

of an object is its ability to do things and to cause change (*work*). Energy can be transferred or transmitted (moved) and can be changed (transformed) into another form. See **energy sources; energy types; energy transfer; energy transformation**.

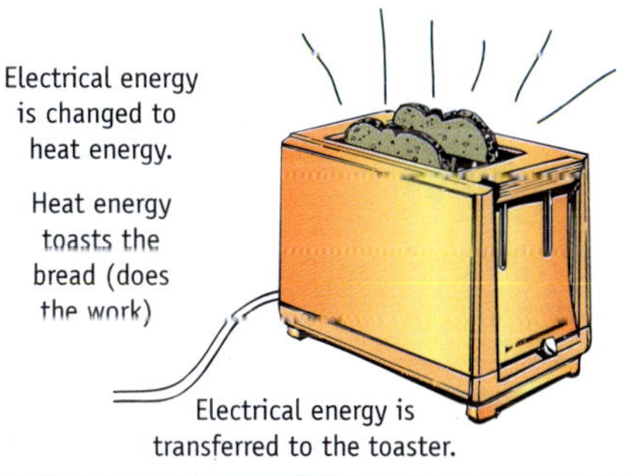

energy conservation

See **conservation of energy**.

energy sources

used by early civilisations for fuel were wood, charcoal and oil (from *plant*s and seeping from the ground). In the 19th and 20th centuries oil, *gas*, *electricity* and nuclear power techniques were developed. See **energy sources, renewable**; **energy sources, non-renewable**.

energy sources, non-renewable

include *fossil fuel*s (coal, oil, natural *gas*). They are used to produce *electricity* and provide petrol and other fuels. *Nuclear energy* obtained from *uranium* metal is also used to generate electricity. Compare **energy sources, renewable**. See **non-renewable resources**.

energy sources, renewable

include *hydroelectricity*, *wave energy*, *tidal energy*, *solar energy*, *wind energy* and *geothermal energy*. They can be used to help conserve *fossil fuel*s. Compare **energy sources, non-renewable.** See **renewable resources.**

energy transfer

occurs when *energy* is passed on from one object to another. For example, vibrations from an *earthquake* travel through the ground and *air* and can be felt in distant places. See **echo; energy transfer in living things**.

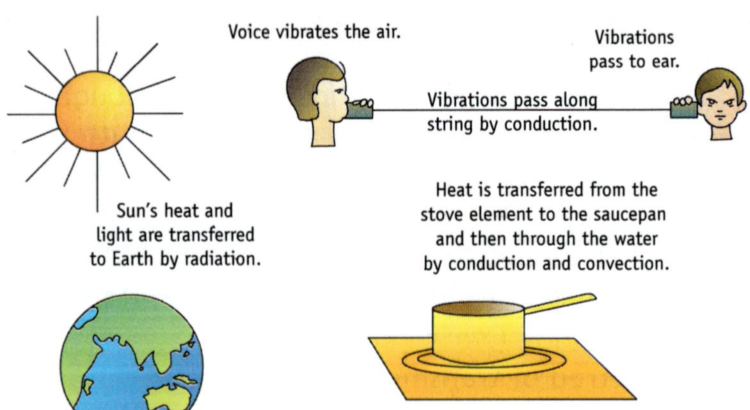

energy transfer in living things

is passing *energy* from one *living thing* to another. It occurs when *food* is passed along a *food chain*.

energy transformation

is when *energy* is changed into a different form. No energy is lost.
See **energy transformation in living things**.

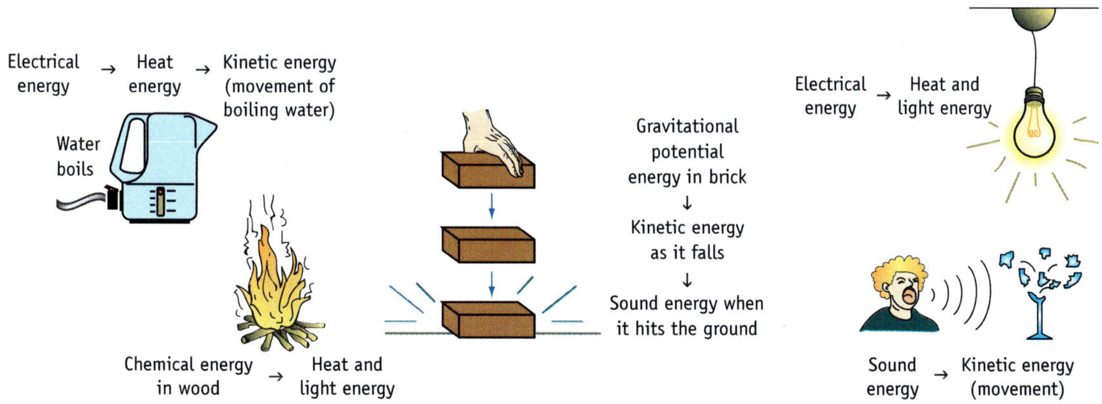

energy transformation in living things

is when one form of *energy* is changed into another. *Solar energy* is converted to *chemical energy* by *plant*s during *photosynthesis*. *Organism*s convert the chemical energy stored in *food* to other forms, such as *movement energy* and *heat energy*.
See **energy used in human activities**.

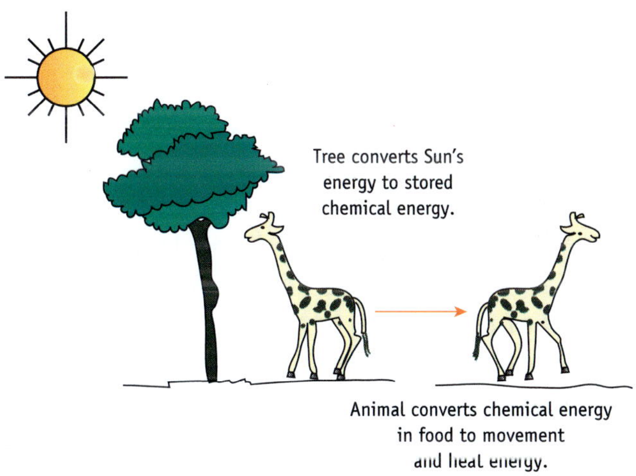

energy transmission

See **energy transfer**.

energy types

Energy exists in many different forms including movement, *heat*, *light*, *sound*, *electrical*, *chemical* and *nuclear*. It may be grouped as *kinetic* or *potential energy*. See **solar energy; radiant energy; energy sources**.

energy we use

comes mostly from oil (petrol) and coal. Oil and Petrol are used mainly for transport. Coal is used to produce electricity and gas. Uranium, a source of nuclear energy, is sometimes used to produce electricity. Electricity is also produced from Coal. Some renewable forms of energy are hydroelectricity, wind energy and solar energy.

energy used in human activities

depends on the size and age of the person, as different activities require different amounts of *energy*.

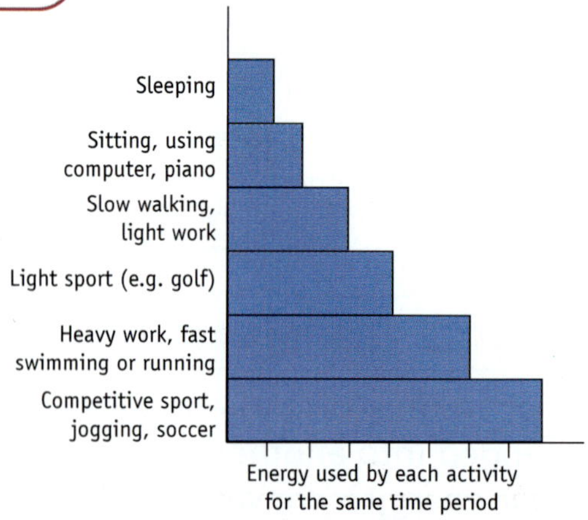

environment

is an *organism*'s surroundings. It includes *living things* (*plants*, *animals*, *bacteria* and *fungi*) as well as non-living things such as *rocks*, *air*, moisture, *temperature* and *light*. Changes in the environment may affect *interactions in the ecosystem*. See **terrestrial; aquatic; environmental changes**.

environmental changes

may alter the interactions in the *ecosystem*. *Drought*, *bushfires and floods*, as well as humans, change the environment. Activities of humans, such as removing forests, *building*, *mining*, polluting and introducing new *species* may result in changes that cannot be reversed. See **ecosystem interactions**.

enzyme

is a chemical that assists in a *chemical reaction* but does not take part in it. Enzymes are in saliva, digestive juices made by the stomach and intestine wall, and the pancreas. They are important in chemical *digestion*, as well as *respiration* and *photosynthesis*. See **catalyst**; **digestion, chemical**.

erosion

is the wearing away of the land. *Rocks* are *weathered* and the rock fragments formed are carried to another place where they are deposited. See **erosion from water; erosion from wind; erosion from moving ice; soil erosion**.

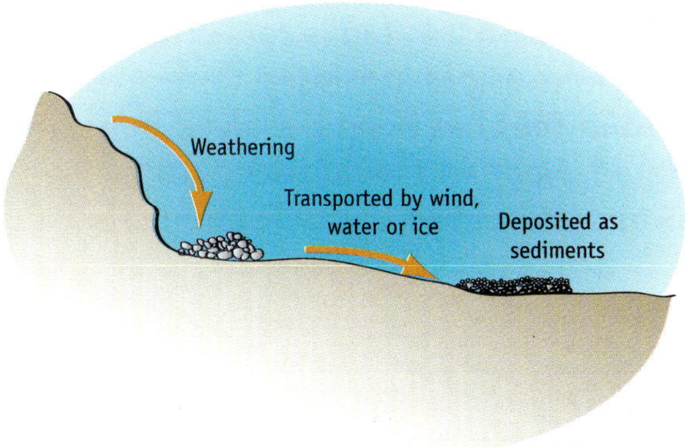

erosion from moving ice

occurs when moving ice carries stones that grind into the valley, wearing down the land. A *glacier* is a large mass of ice slowly moving down a valley. When the ice *melts*, the stones are deposited. See **corrasion; glacial landforms.**

erosion from water

occurs because rivers and ocean waves often carry rock fragments and sand. These particles have an abrasive effect and grind into the riverbeds and cliff faces. When the water slows, the particles are deposited. See **corrasion; river landforms; coastal landforms**.

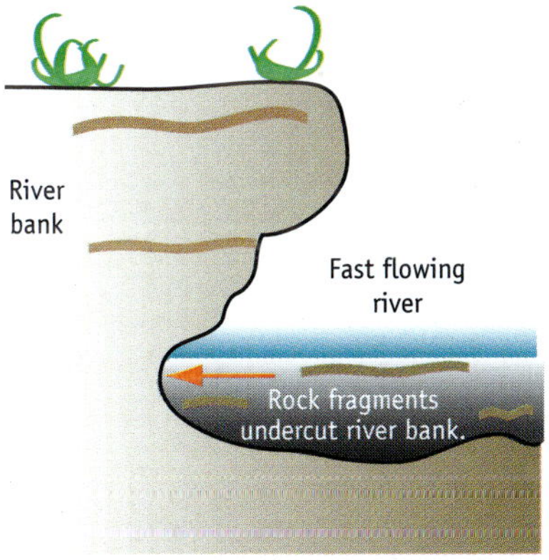

erosion from wind

occurs because strong winds often carry *weathered rock* particles (e.g. sand). The particles have an abrasive effect and wear down the surrounding landscape. As the *wind* slows down, the particles are deposited. See **corrasion; desert landforms**.

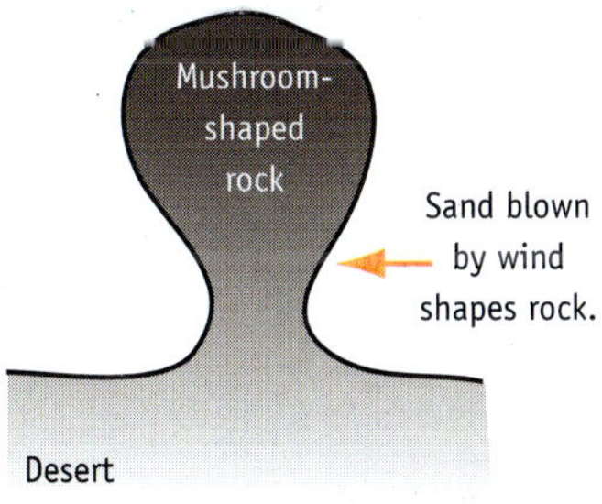

evaporation

occurs when a *liquid* is converted to a *vapour* or *gas*. It is used to recover a *solid* that has been *dissolve*d in a liquid. *Heat* and *wind* speed up evaporation. See **crystallisation**.

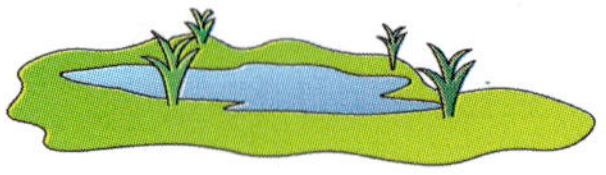

Water evaporates slowly from the puddle and it eventually dries up.

Water evaporates quickly from the kettle as the water boils.

evergreen

describes a tree or shrub that keeps its leaves throughout the *year*. Compare **deciduous**.

evolution

is a series of gradual changes in *organism*s and the development of new varieties and *species* over a long period of time. See **fossils**.

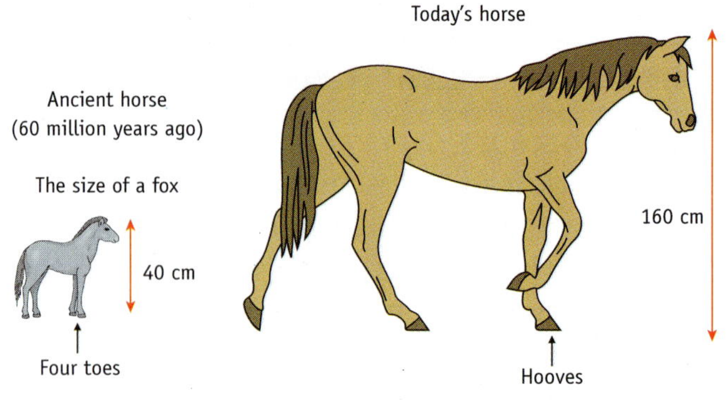

excretion

is the removal of wastes that have formed in *cell*s during chemical reactions. Wastes are removed by the kidneys and bladder as urine and by the skin as perspiration. *Carbon dioxide* is removed by the lungs. See **excretory system; respiratory system**.

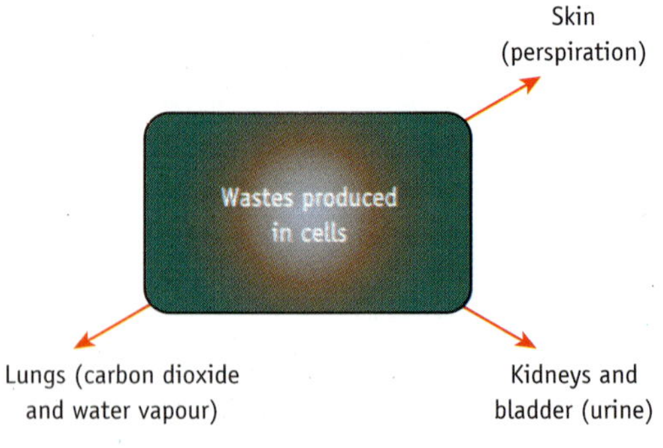

excretory system

consists of kidneys, ureters, bladder and urethra. Chemical wastes from *cells* are taken by the blood to the excretory system, where they are removed from the body. The lungs and the skin are also excretory *organs*. See **respiratory system**.

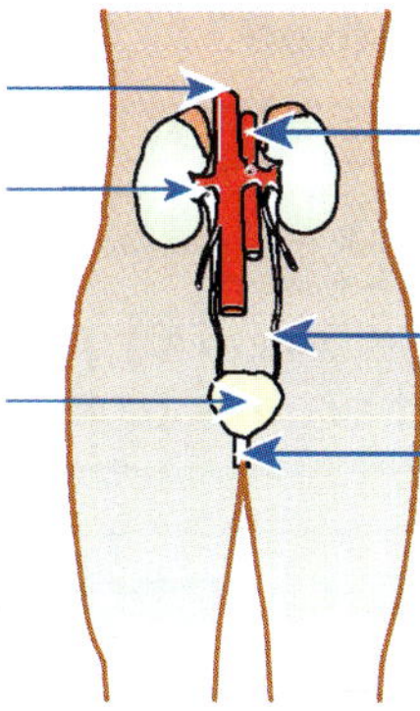

expansion of matter

occurs when the *matter* is heated. Heating makes the particles move more and spread out, taking up more space. See **particle theory of matter**. Compare **contraction**.

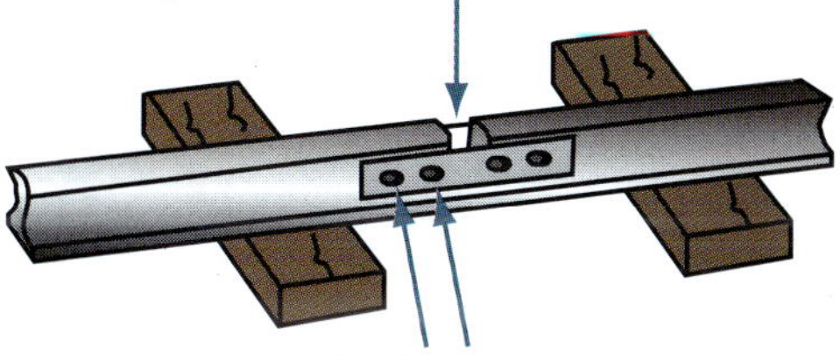

experiment

See **scientific method**.

extinction

is when *organisms* disappear from the *Earth* because of changes in the *environment* (e.g. *climate*) and human activities (e.g. land clearing and introducing other *animals*). The Tasmanian wolf or 'tiger' (thylacine) has not been seen since the early 1930s. See **megafauna**.

Tasmanian wolf

fair test

See **scientific method**.

fats and oils

are organic *nutrients*. They provide an *energy* store and an *insulat*ing layer under the skin and around *organs* to keep the body and organs warm. See **organic compound**.

feedback and feed forward systems

are *systems* the *organ*s use to send or 'feed' messages to the brain, to keep the body working properly.

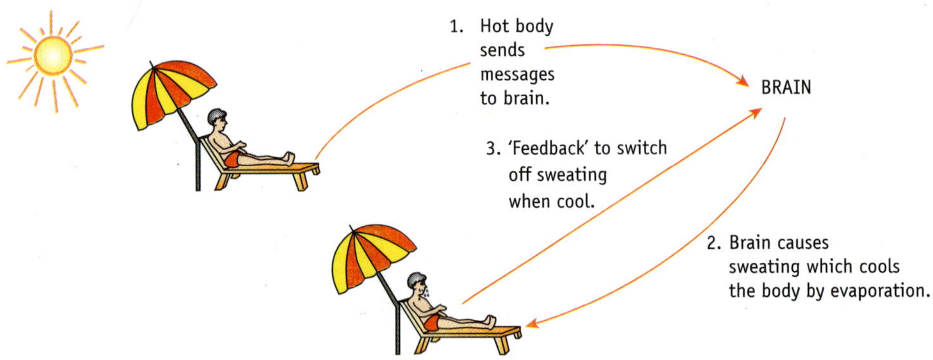

feeding relationships

See **food chains; predation; parasitism.**

feral

refers to *plant*s and *animal*s that are wild but not in their native *habitat* (e.g. feral cats). They compete with or *prey* on native *organism*s.

fertilisation

is the joining of *egg* (ovum) and *sperm* during *sexual reproduction*. The fertilised egg grows and develops to form a new individual similar to its parents. See **reproductive adaptations.**

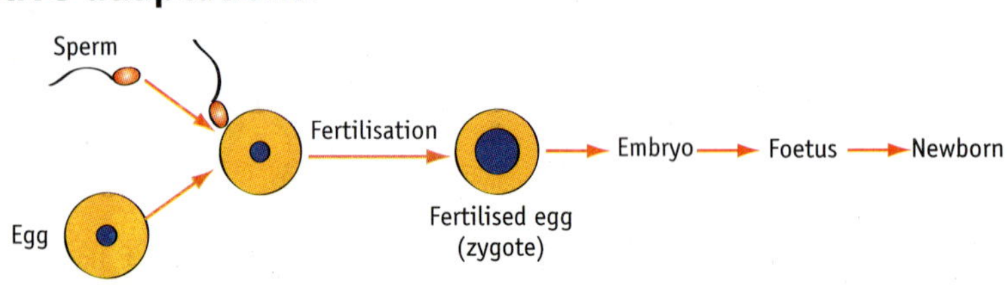

fibres

are threads that make up fabrics and ropes. Cotton, wool, linen (from flax) and silk are natural fibres. Nylon, polyester and acrylics are made, or *synthetic*, products. Most synthetic fibres are made from *petroleum*.

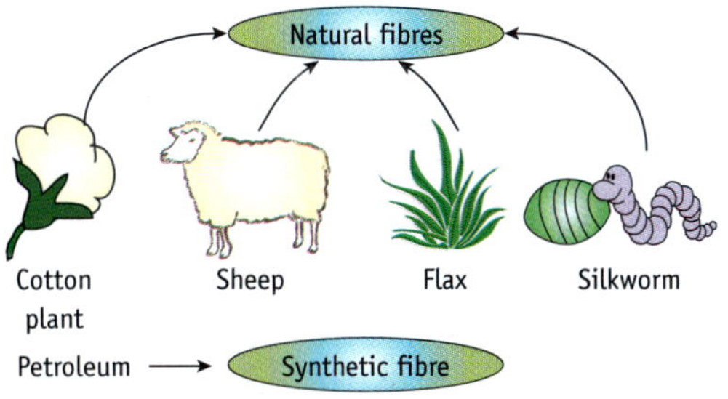

field

is an area in which a *force* acts from a distance. See **magnetic field; electrostatic (electric) field**.

filtration

is used to separate particles from a *liquid* or *solution*. For example, when muddy salt water is filtered, the mud is collected on the filter and the salt water passes through it. Filtration is used to remove dirt from a carpet, and dust or grit from petrol, air and oil to prevent the grit and particles from contaminating drinking and swimming pool water. It is also used to prevent grit passing into the engine of a car.

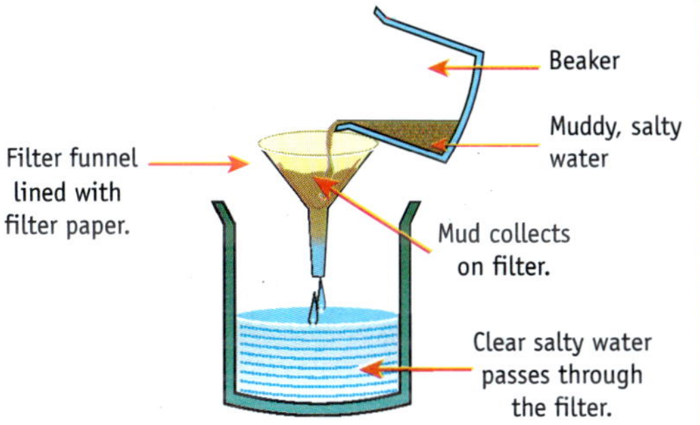

float

means to stay on the surface of a fluid. An object will float if its *weight* and the upward *force* (or *upthrust*) are equal. An object light for its size will float. Its *density* is less than the density of the surrounding fluid. See **sink; buoyancy**.

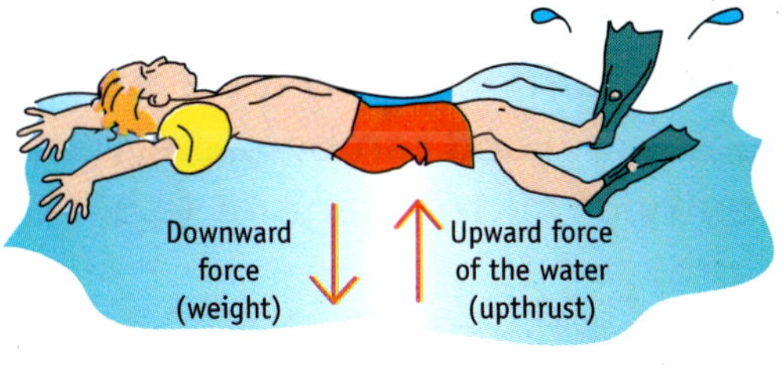

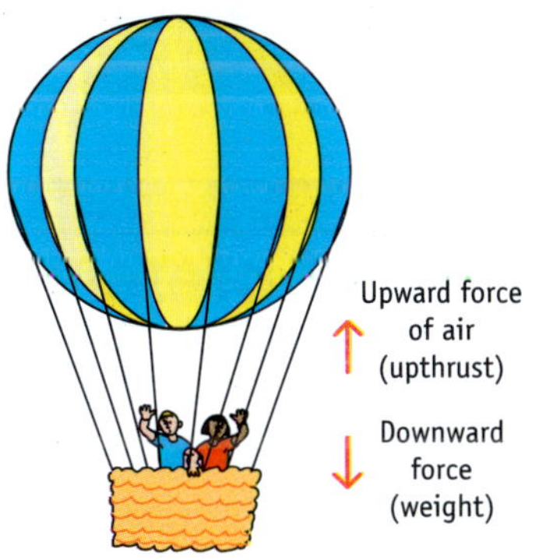

flower

is a *plant structure* in which *sexual reproduction* occurs. The anthers make pollen, which contains the male sex *cells*. The ovary makes female sex cells or *eggs*.

See **reproduction in flowering plants**.

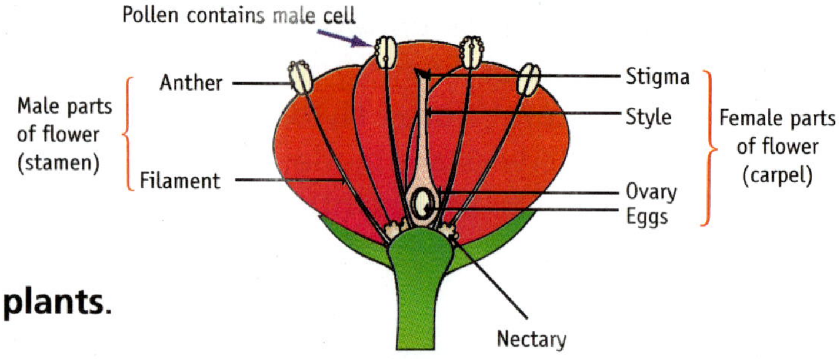

flowering plants

have *flowers*, and produce *seeds* and *fruit*. See **plant structure; life cycle of a flowering plant**.

food

See **nutrients**.

food chain

shows the feeding relationships in an *ecosystem*. Each food chain starts with a *producer* (plant), which is eaten by a *consumer* (*animal*). The arrows show the direction of flow of *nutrients* and *energy* in the ecosystem. See **food web**.

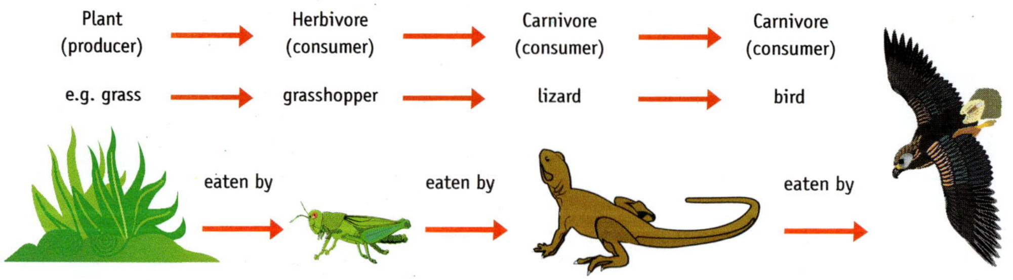

food preservation

is when the *chemical reactions* of *microorganisms* in *food* are stopped. These *chemical reactions* cause decomposition. Food can be preserved by:

- killing and/or preventing the entry of microorganisms (e.g. *vacuum* sealing, salting, smoking, irradiating), or
- slowing or stopping the chemical reactions by freezing or drying.

food pyramid

(e.g. pyramid of numbers) can be used to illustrate a *food chain*, with the *producers* at the base and then *herbivores* and *carnivores* as they occur in the food chain.

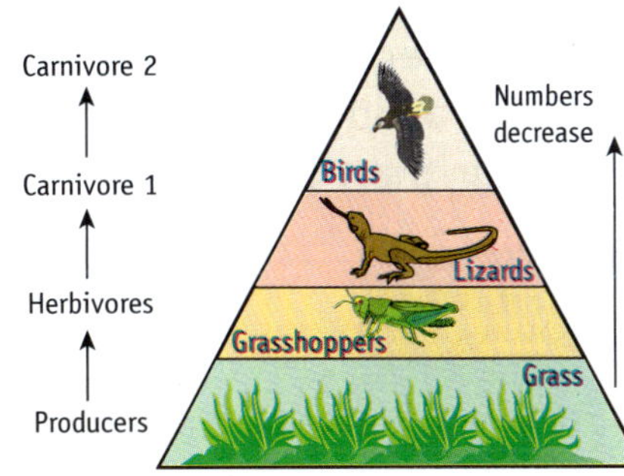

food web

shows feeding relationships in an *ecosystem*. In nature many *food chains* combine to form a complex food web. See **food chain**.

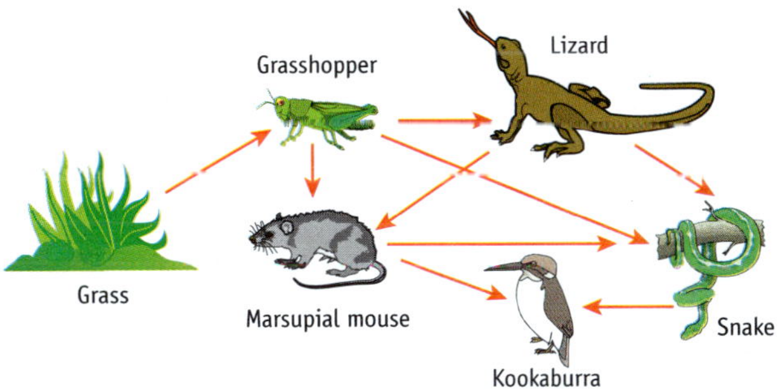

force

is a push or a pull.

A force is needed to start an object moving or to *speed* its movement.

A force is needed to slow down or stop movement.

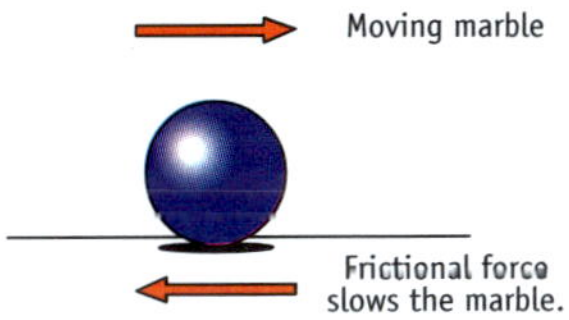

A force is needed to change the direction of movement.

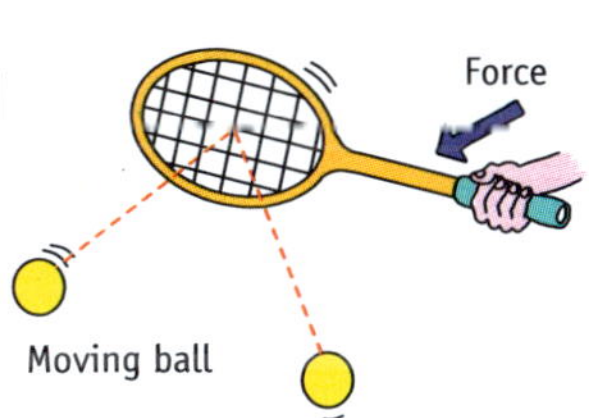

A force is needed to change shape.

See **forces, balanced and unbalanced; gravity; electrostatic force; magnetic force.**

forces, balanced and unbalanced

determine whether an object will move. When an object is not moving (is at rest), the *force*s acting on it are balanced. To make an object move, one force must be greater than the others—the forces must be unbalanced.

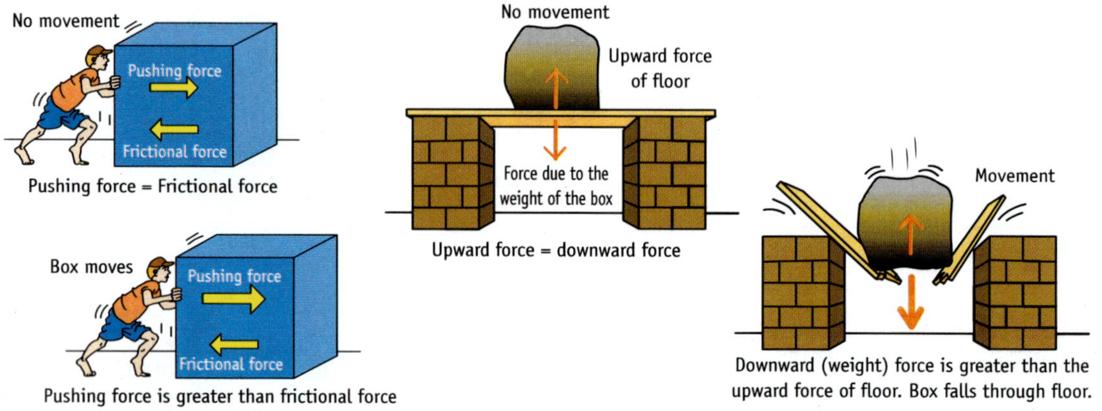

forces between particles in matter

hold the particles together. These forces of attraction are called *bond*s and they give a substance its *properties* (e.g. *boiling point* and hardness).

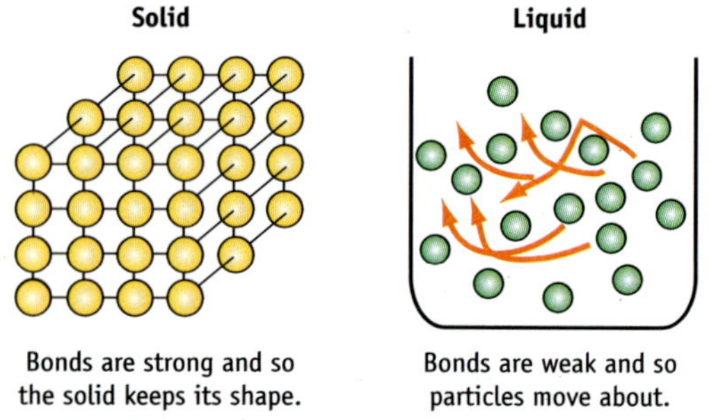

fossil

is a remain or trace (e.g. bone, footprint) of ancient life. Fossils show us what life in the past was like. For example, archaeopteryx is a fossil that has some bird and some reptile features.

See **dinosaurs; fossil fuel; rock age**.

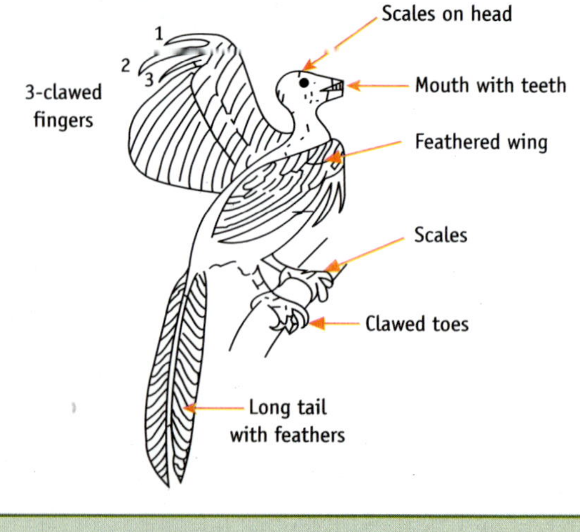

fossil fuel

is obtained from the *Earth* and was made from *plants* or *animals* millions of years ago. *Coal*, *petroleum* and natural *gas* are *fossil* fuels. They are *non-renewable resources*. See **fossil fuel use; conservation of non-renewable resources**.

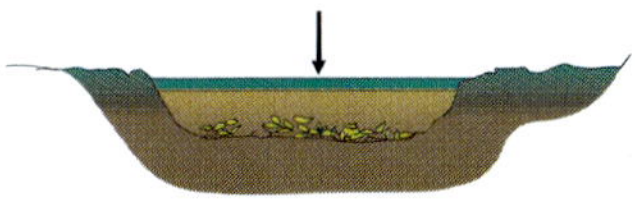

Sediments are compressed by layers of soil and so form peat.

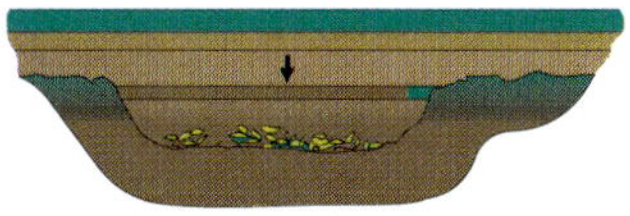

More compression changes the peat to coal.

fossil fuel use

provides about 90% of the world's *energy*. Humans use these fuels for heating, cooling and cooking, and in industry. Once used, these fuels cannot be replaced. See **petroleum**.

Products from petroleum:

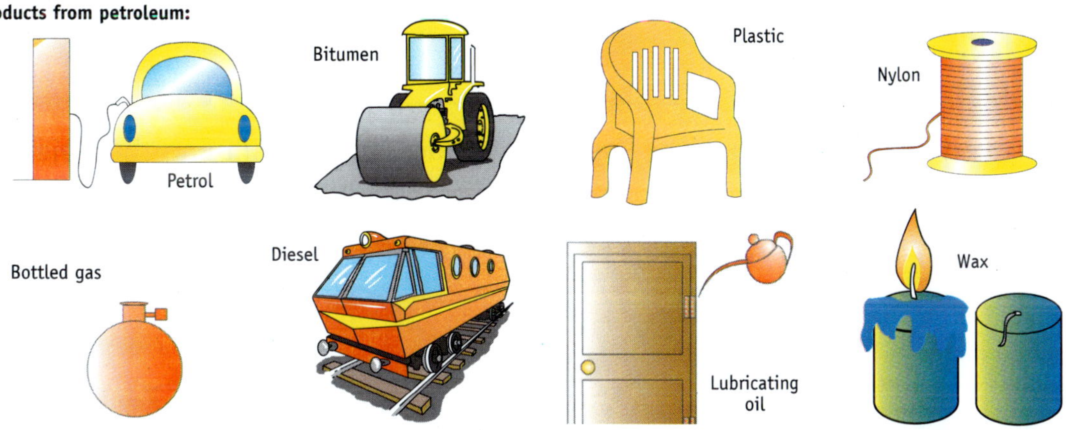

foundations

are the base on which structures are built. Before excavation or constructing buildings, roads, tunnels, *bridges* or *dams*, engineers study the *rocks* and *soil* of the area. Soft rock or cracks may cause the structure to sink or collapse. Clay swells when it is wet and shrinks when dry; this causes cracking in buildings.

freezing point

is the *temperature* at which a *liquid* turns into a *solid*. Different substances have different freezing points, but it is always the same as the *melting point* of the substance.

frequency

of a wave is the number of waves that pass in a second. See **pitch of a sound**.

friction

is a *force* that tries to slow or stop the movement between surfaces. Friction is less between smooth or slippery surfaces. See **friction reduction**.

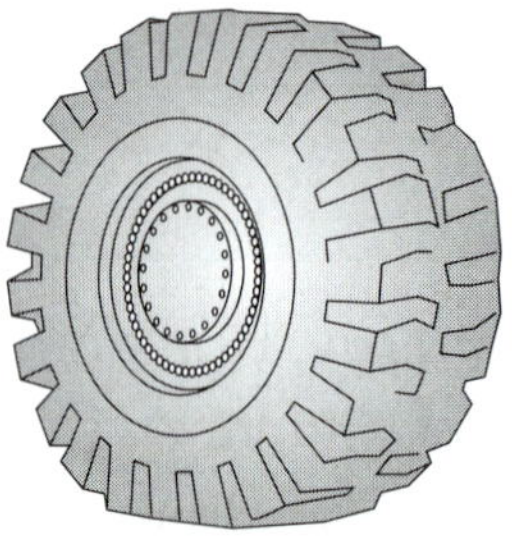

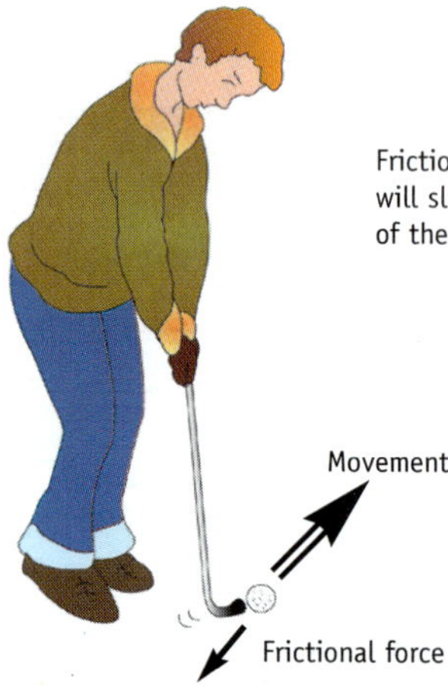

Friction: the golf ball will slow down because of the frictional force.

friction reduction

is achieved by oiling, sanding or waxing a surface, or streamlining the object.

Skiers put wax on the skis to reduce friction.

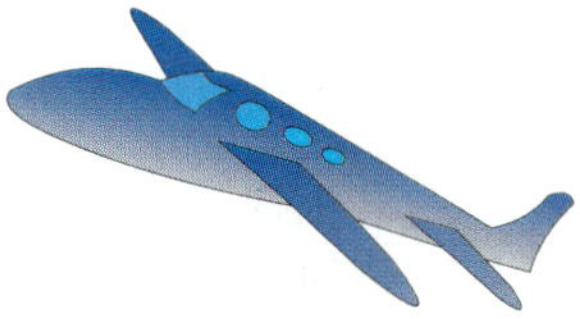

Aircraft and racing cars are designed to reduce friction.

Animals are streamlined to reduce friction.

Oil is used to reduce friction in machinery.

fruit

may be fleshy (e.g. peach or tomato), woody (e.g. gumnut), a dry capsule (e.g. poppy) or a dry pod (e.g. pea pod). A fruit encloses *seeds*.

Tomato

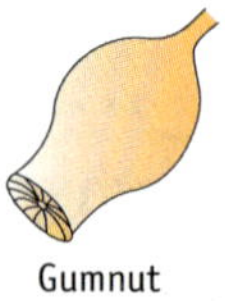

Gumnut

Poppy capsule

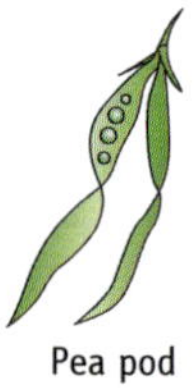

Pea pod

fuel

is a source of *energy*. See **energy source; fossil fuel**.

fungus (plural: fungi)

is a *decomposer*. The fungus *kingdom* includes mushrooms, yeast, mould, and rusts and smuts that cause disease in *plants*. Fungi have plant-like *cells*, but no *chloroplasts*. See **microorganisms**.

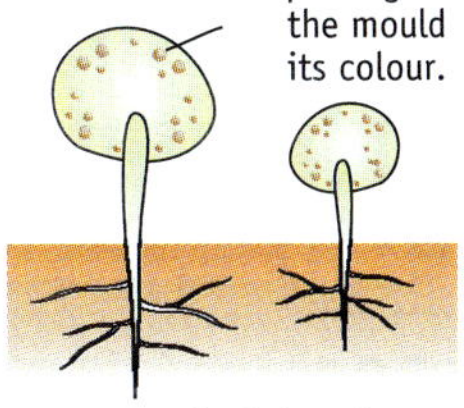

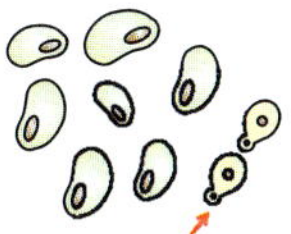

galaxy

is a huge cluster of *stars*. There are millions of galaxies moving around the *Universe*. Our galaxy is a rotating spiral made up of billions of stars, dust and *gases*. The band of stars seen across the sky is part of our galaxy and is called the *Milky Way*.

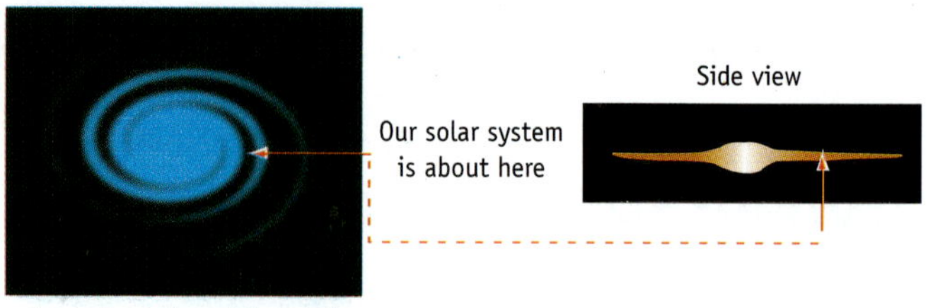

Our galaxy viewed from below.

galvanise

See **rustproofing**.

gas

is one of the *states of matter*. Gases have no definite size and they spread out to fill a container and take its shape. See **particle theory of matter**.

gears

are simple *machines* made of wheels with teeth (cogs). When one wheel turns, it transfers its motion and *energy* to another wheel. Gears are found in bikes, clocks and many machines.

a) Gear wheels that are the same size change the direction of movement.
b) A small gear wheel will make a large wheel exert more *force*.
c) A large gear wheel turning slowly will make a small wheel rotate fast.

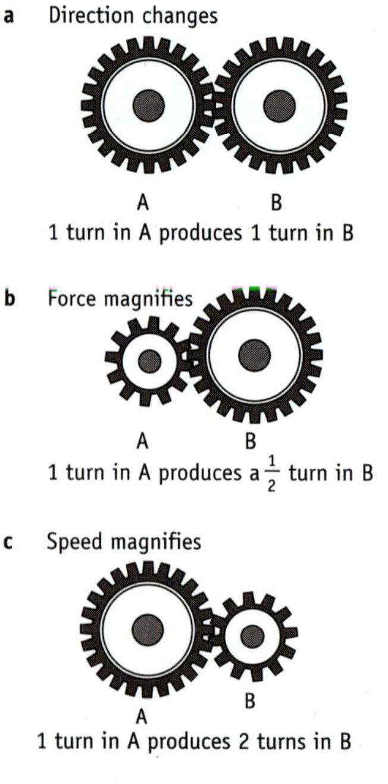

generator

is a *machine* that produces an *electric current*. It changes *kinetic energy* to *electrical energy*, usually by rotating a coil in a *magnetic field*.

genetic engineering

is the process of altering genes present in an *organism*. Scientists have separated useful genes and transferred them to other organisms.

Cotton plants can be given a gene that makes them resistant to caterpillars.

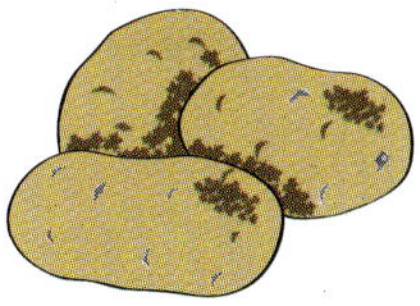

Potato plants can be given a gene that helps them resist virus attack.

Tomato plants can be given a gene to stop them squashing.

geothermal energy

comes from *heat* in underground water and *rocks* in volcanic areas. It is brought to the surface as steam. The steam turns *turbines*, which drive *generators* and produce *electricity*. See **energy sources, non-renewable**.

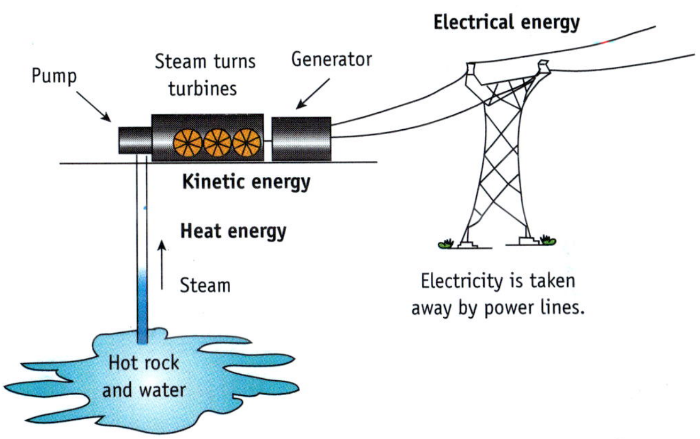

germination

occurs when the small *plant* (embryo) inside a *seed* begins to grow. Seeds need water, moisture and warmth. Some seeds (e.g. wattles) germinate better after fire.

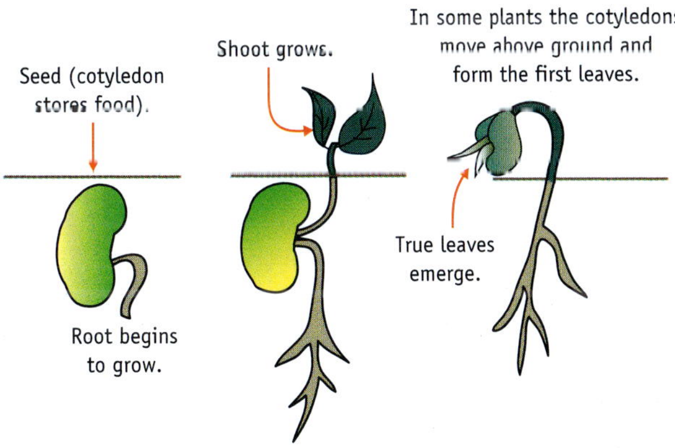

G

glacial landforms

occur in mountainous areas and have U-shaped valleys, hanging valleys and waterfalls. *Rock* deposits are found along the sides and ends of the valley.
See **erosion from moving ice**.

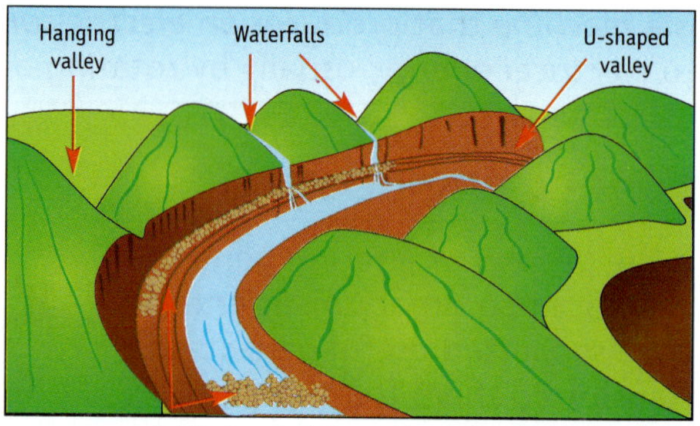

Rubble (moraine) dropped by glacier when it melted

glacier

is a mass of ice slowly moving down a mountainside or off the edges of a polar ice-cap. See **erosion from moving ice**.

global air currents

occur when *air* moves along the *Earth*'s surface from the poles towards the equator. The hot ground at the equator heats the *air* causing it to rise. This warm air moves above the ground, back to the poles. The air currents curve to the west because of the Earth's rotation.
See **convection in the atmosphere**.

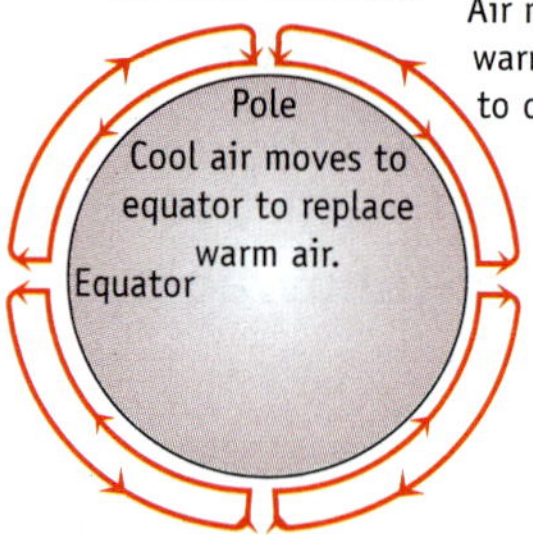

global ocean currents

occur in the surface water of the oceans. They are caused by *wind* circulation, *convection* and the *Earth*'s rotation. The *Sun* heats the oceans and currents carry this *heat* to other places. This affects the *weather* and *climate* around the Earth.
See **El Niño.**

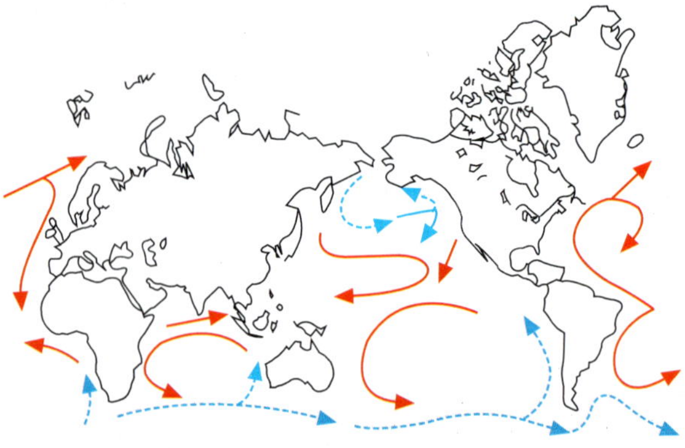

global warming

is the increase in the *Earth*'s *temperature*s as a result of excess *carbon dioxide* in the *air*. Carbon dioxide has increased because of overuse of *fossil fuel*s and the removal of forests (trees remove carbon dioxide from the air). Pollutants, such as fluorocarbons from aerosol cans and refrigerants, also trap *heat*. See **greenhouse effect**.

Greenhouse gases trap heat and increase global warming.

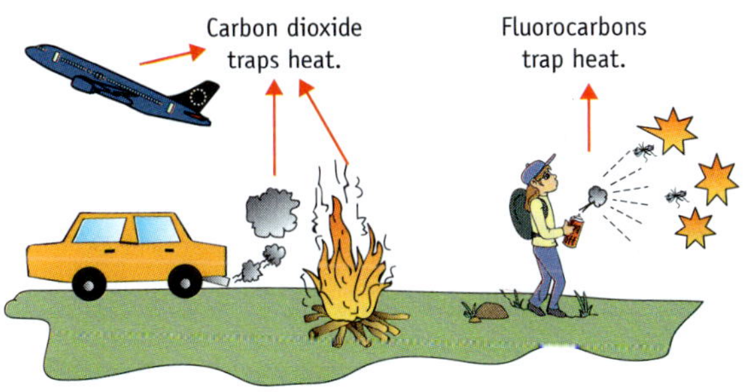

Destroying plants increases global warming.

granite

See **igneous rocks**.

gravitational force

See **gravity**.

gravity

is a *force* of attraction that exists between all *matter* in the *Universe*. The *Earth* has a gravitational force that pulls things towards its centre and makes them fall. This force is the object's *weight*. See **gravity and mass; gravity and distance**.

Gravity pulls the ball back to the ground.

gravity and distance

are connected. *Gravitational force* becomes less when objects are further apart.

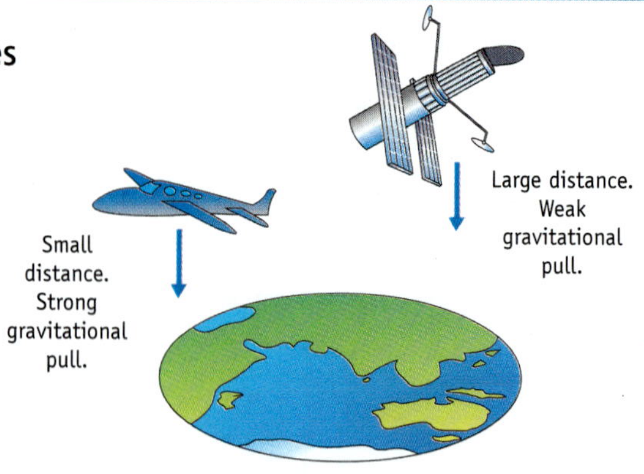

gravity and mass

are connected. Objects with large *mass* (that contain a large amount of *matter*) have a greater gravitational *force* than objects with small mass. For example, the *Earth*'s pull of *gravity* on an object is greater than the *Moon*'s pull on the same object. See **weight**.

greenhouse effect

describes how *carbon dioxide* in the *atmosphere* traps *heat* that is radiated from the *Earth*'s surface. This keeps the outer surface of the Earth warm in a similar way to the glass roof of a plant greenhouse.

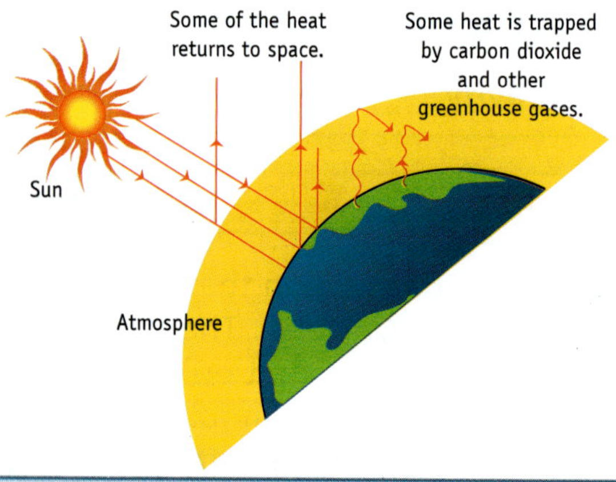

habitat

is the place where an *organism* lives (e.g. creek, bushland or desert). *Plant*s and *animal*s have features or *adaptations* to help them survive in their habitat. See **aquatic; terrestrial**.

Adaptation for aquatic habitat

Adaptation for a dry/hot habitat

Australian grass tree

heat

See **heat energy**.

heat absorption

is how objects take in *heat* from warmer surroundings. Dark objects absorb heat more rapidly than light-coloured objects.
Compare **heat radiation**.

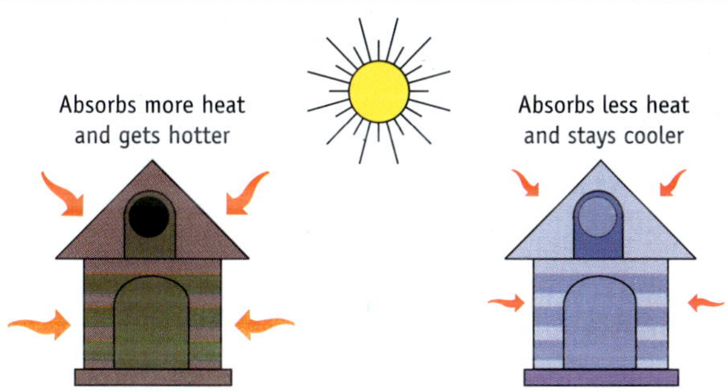

heat energy

makes things hot. It makes the particles in a substance vibrate more and move apart so that the object expands. Heat is detected by a *thermometer* and is measured in degrees Celsius (°C). See **heat movement; energy transfer; energy transformation**.

heat movement

is when *heat* moves from a hot place to a cooler place. It can be moved or transferred from one place to another by *conduction*, *convection*, *radiation* and *reflection*. See **thermos flask**.

heat radiation

is one way *heat energy* can travel. The heat travels in beams or rays (infra-red rays). This is *radiant energy* and does not need particles to travel; it can pass through a *vacuum*. Compare **conduction of heat; convection**.

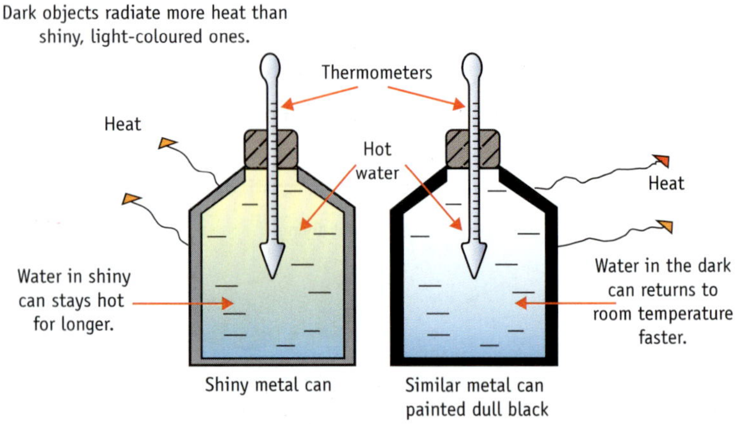

heavy metal

is *metal* made up of large particles. Some of these metals (e.g. lead and mercury) are poisonous and can pollute the *environment*. See **waste disposal**.

herbivore

is an *animal* that eats *plants* (e.g. kangaroo, cow). See **food chain**.

hibernate

is what some *animals* (e.g. bats) do to conserve *energy* when *food* is scarce. During winter they become inactive and their heart rate and *breathing* become very slow. *Carnivores* such as bears do not actually hibernate but may sleep for long periods of time. See **nutrients**.

high pressure system

often brings calm, fine *weather*. In the southern hemisphere, the *air* moves away from an area of high *air pressure* in an anticlockwise direction. Compare **low pressure system**. See **wind**.

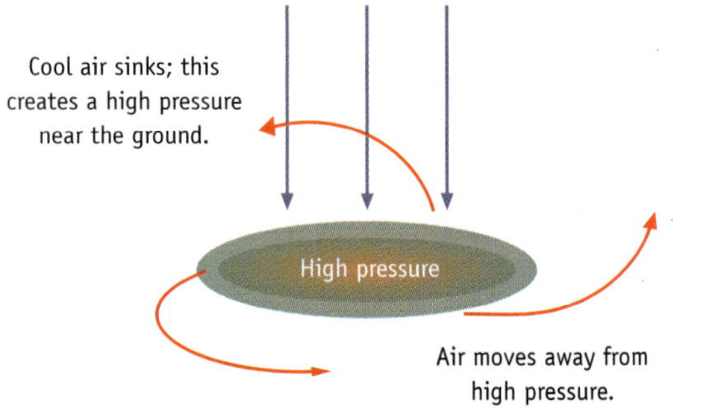

Hubble Space Telescope (1990)

was designed to see into outer *space* while it is in *orbit*. It operates outside the *Earth*'s *atmosphere* and so produces images that are not distorted by the atmosphere.
See **telescope**.

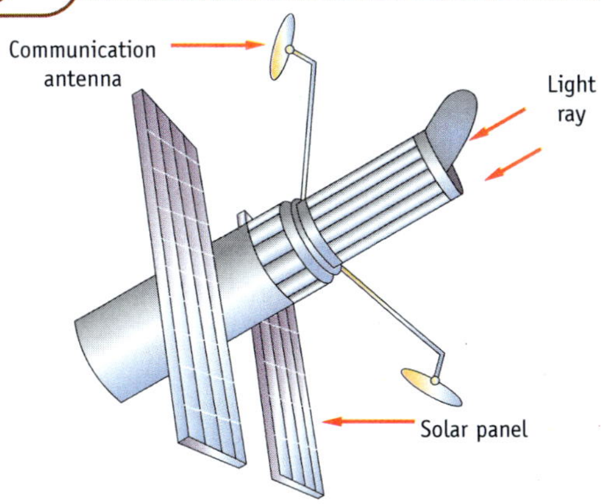

humidity

is the water *vapour* or moisture in the *air*. As air cools, some of the water vapour condenses to form *liquid* water. This produces *clouds*, fog and dew. Humidity is measured by a wet-and-dry-bulb *thermometer*.

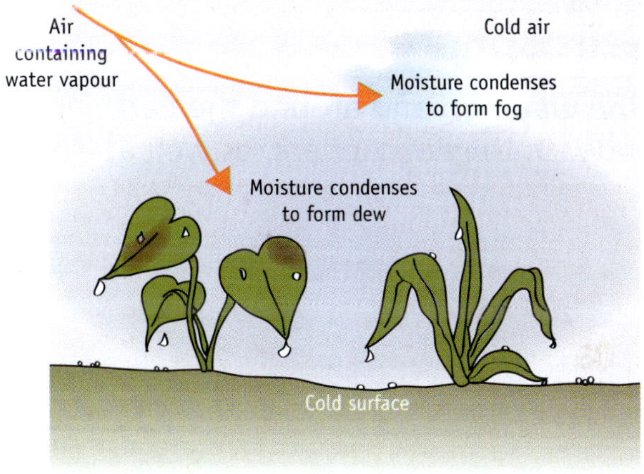

hydraulic machine (system)

transfers *energy* by applying *pressure* on a *liquid*. Pressure is transferred equally throughout the liquid.

$$\text{Pressure} = \frac{\textit{force}}{\text{area}}$$

This system is used to lift heavy *weights*, such as hoists in car repair shops.

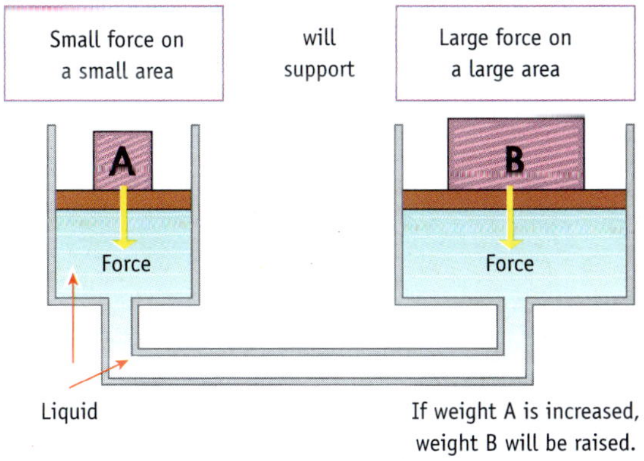

hydroelectricity

is *electrical energy* produced by *generator*s driven by flowing water. The flowing water is commonly provided by *dams* built in mountainous areas. See **energy sources, non-renewable; wave energy; tidal energy**.

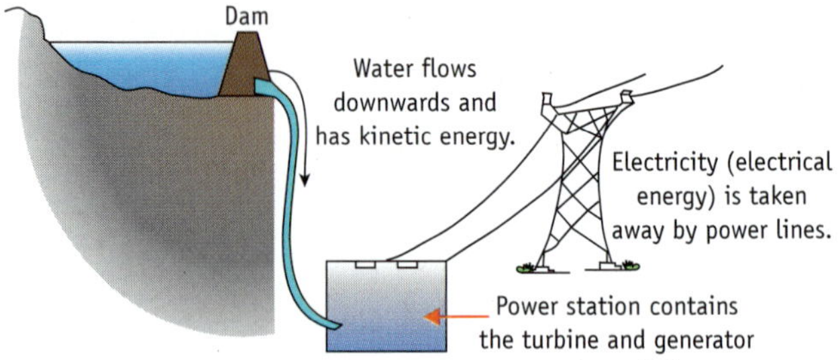

hydrosphere

is the water on and around the *Earth*, including seas, oceans, moisture and *vapour* in the *atmosphere*, as well as ice and snow. Compare **lithosphere**.

igneous rocks

are formed deep inside the *Earth* from *magma* (molten or melted substances) or on the surface from *lava* from volcanoes. As the molten rock cools, it forms interlocking *crystal*s. See **crystal formation**.

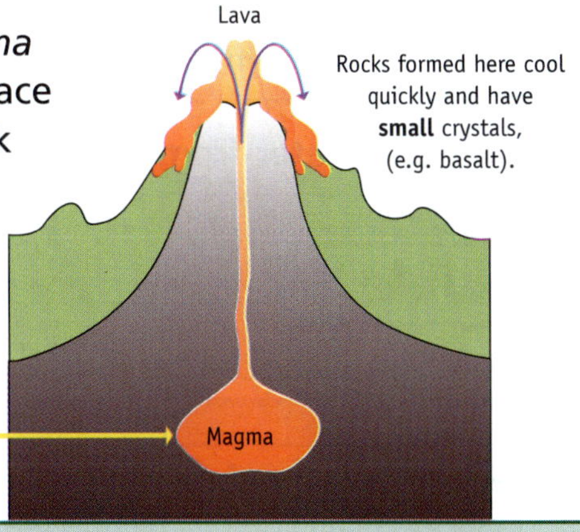

impact

describes the effects human activities and natural disasters have on the *environment* and *Earth*. The impact may be helpful and/or harmful (e.g. the use of *fossil fuel*s provides *energy* but also pollutes and uses *non-renewable resources*).

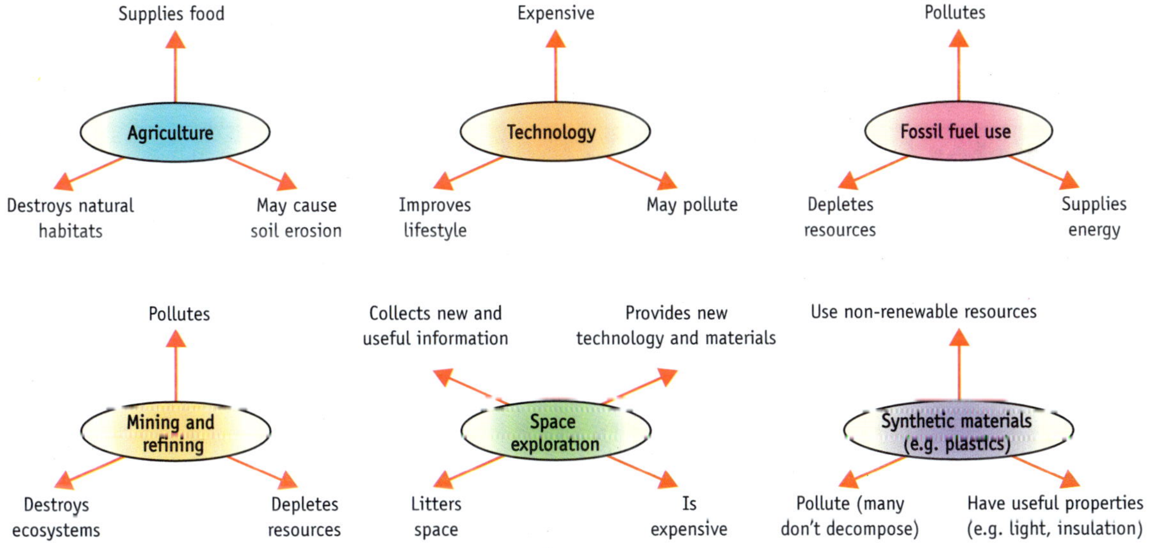

inclined plane

See **ramp**.

indicators

are used to detect *acid*s. See **pH**.

inorganic nutrients

are *minerals* and water that are needed for growth and *chemical reactions* in *living things*. *Animals* obtain inorganic *nutrients* from their *food*. *Plants* obtain minerals and water from the *soil* and *carbon dioxide* from the *air*.

insoluble

describes a substance that will not dissolve. Compare **soluble**.

insulator (electricity)

is a poor *conductor* and stops the flow of a *current*. It prevents electric shock and short *circuits*. Electrical wires are usually coated with an insulator such as *plastic* or rubber.

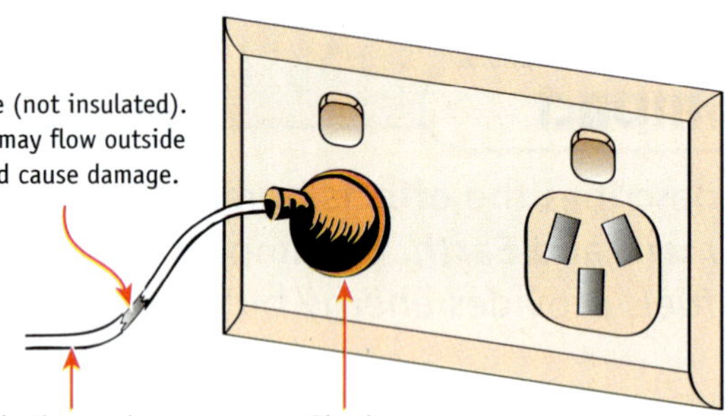

insulator (heat)

is a poor *conductor* and reduces the flow of *heat* from a hot area to a cold one. *Materials* such as wood, *plastic*, glass, rubber and *air* are insulators. They are often used to control movement of *heat*. See **conduction of heat**.

Plastic handle stops heat flowing into hand.

Bird fluffs up its feathers to trap air and stop heat loss.

Foam or polystyrene cups slow down heat loss.

interdependent relationships in ecosystems

See **ecosystem interactions**.

invertebrates

are *animal*s that do not have a backbone. The main groups are shown in the diagram.

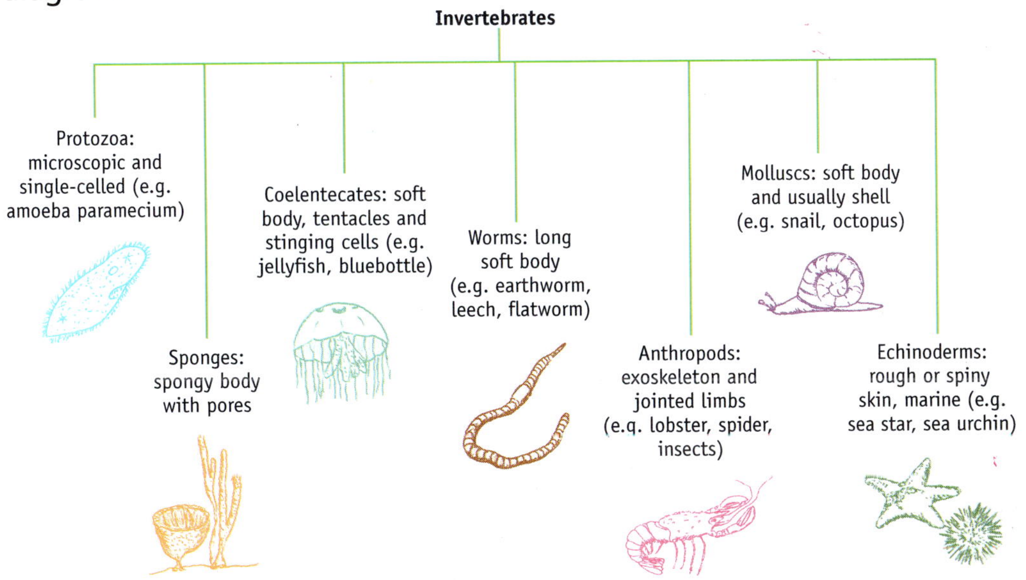

iron ore

See **refining**.

Jupiter

is the largest *planet* and is mainly *gas*. The outer layers show coloured, banded clouds with ovals and spots caused by storms. One storm area, called the Great Red Spot, is three times the size of *Earth*. Jupiter has sixteen *moons*.

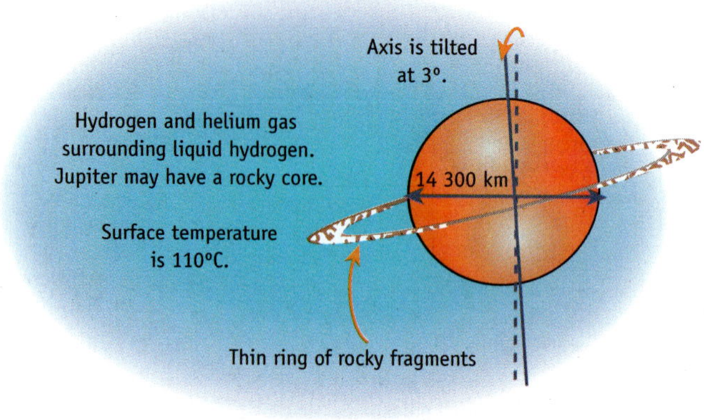

kaleidoscope

is made up of *mirror*s joined at an angle. Coloured paper shapes or beads are placed between the mirrors. *Light* bounces between the mirrors and produces patterns. See **reflection**.

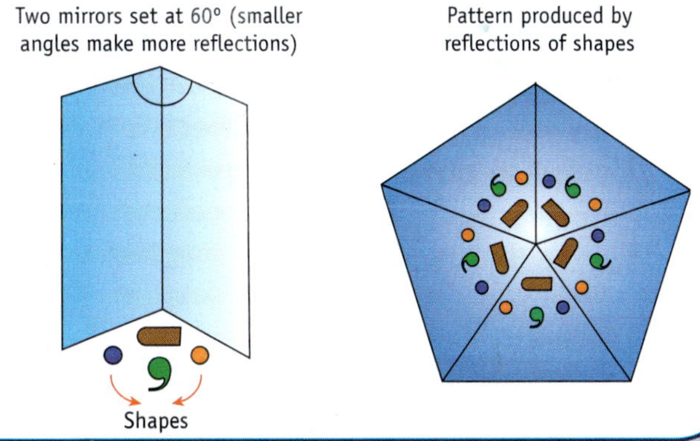

keys

are used to identify and classify *living things*. A key divides or branches into two sections at each step and so is called a dichotomous key (di = two, chotomous = branch). The diagram shows a simple key to identify *animals* with a backbone:

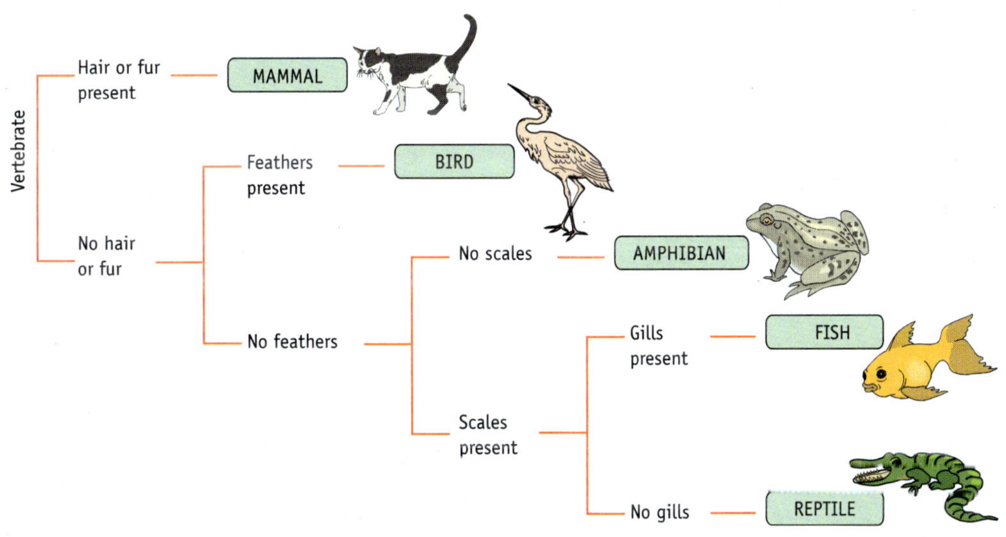

This same key may be written as follows. Select (a) or (b) at each step.

1. (a) hair or fur present → mammal
   (b) no hair or fur → go to 2
2. (a) feathers present → bird
   (b) no feathers → go to 3
3. (a) scales present → go to 4
   (b) no scales present → amphibian
4. (a) gills present → fish
   (b) no gills → reptile

kinetic energy

is the *energy* in a moving object such as *wind*, flowing water, a falling rock, a swinging axe or a moving car. All these are capable of changing an object they hit. The faster an object is moving, the more kinetic energy it has. Compare **potential energy**.

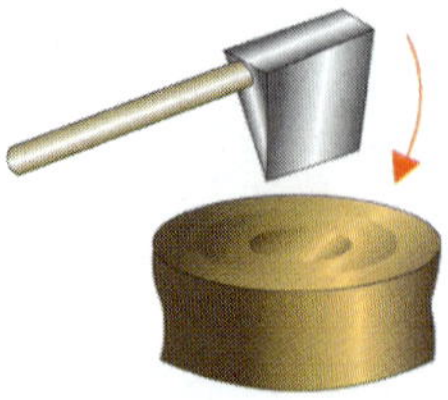

Moving axe has kinetic energy.

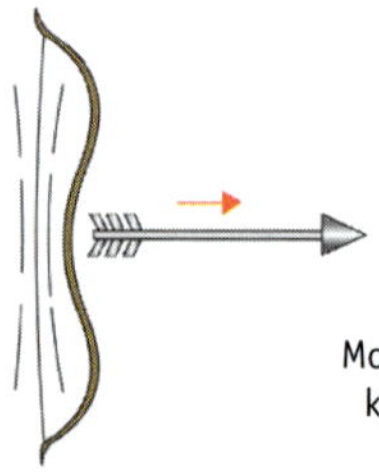

Moving arrow has kinetic energy.

kingdoms

are the first level in the *classification* of *living things*. A five-kingdom system is commonly used. It places all living things into one of five kingdoms: *Plants*, *Animals*, *Fungi*, *Bacteria* and *Viruses*. See **classification of living things**.

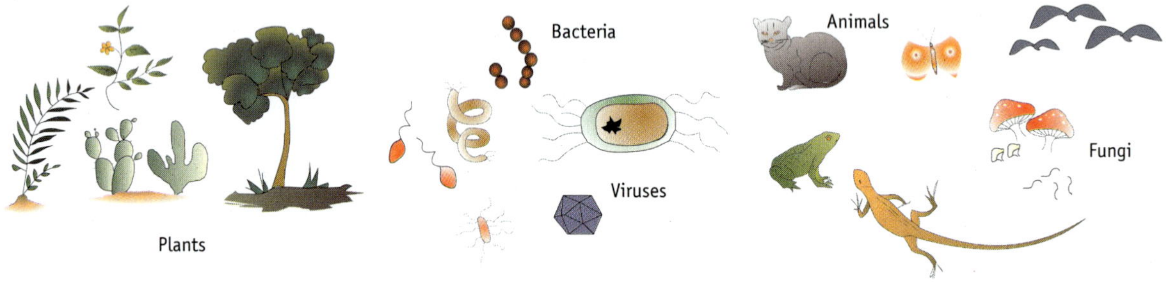

land management

See **resource management**.

landform

is the landscape. It is the natural structures, such as hills, valleys and plains, on the *Earth*. Landforms are formed by *erosion* as well as earth movements (e.g. folding and tilting). See **coastal landforms; river landforms; desert landforms; glacial landforms**.

larva

is the young immature stage of some animals. It hatches from the egg and is unlike the adult in appearance. Examples include a maggot (fly larva) caterpillar (butterfly or moth larva) and tadpole (frog larva). See **metamorphosis; life cycle**.

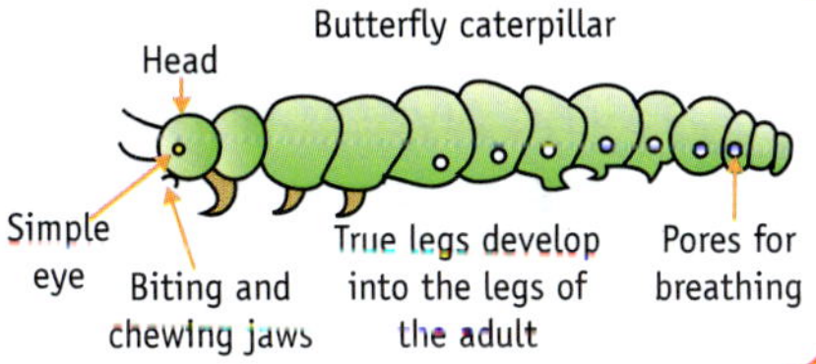

lava

is molten or *liquid rock* that flows over the *Earth*'s surface. See **volcano**.

leaves

make *food* for the *plant* by the process of *photosynthesis*. They transpire and this draws water up the plant. The water cools the plant and carries minerals to *cells*. See **transpiration; stomates**.

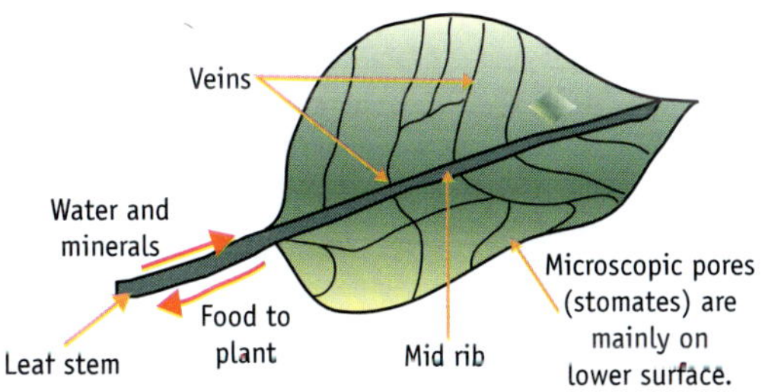

lens

is a disc made of a clear material with one or both surfaces curved (e.g. magnifying glass). Lenses are used in cameras, spectacles, *telescopes* and microscopes. *Light* bends as it passes through a lens. See **refraction**.

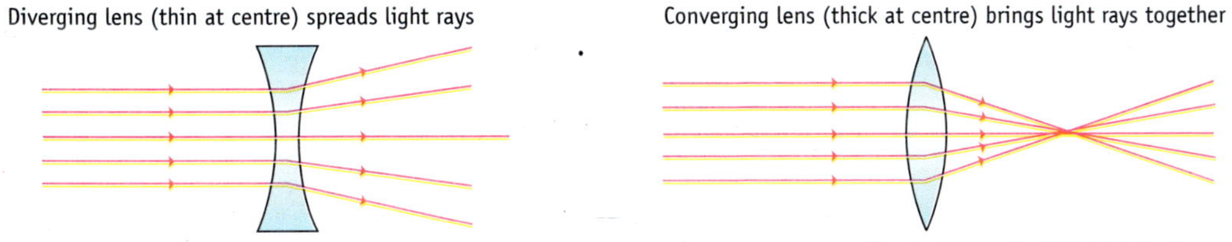

lever

is a simple *machine*. It is a rigid bar that can move about a pivot called a fulcrum. The effort *force* applied at one end of the lever can be used to move a load. See **lever types; levers in the human body**.

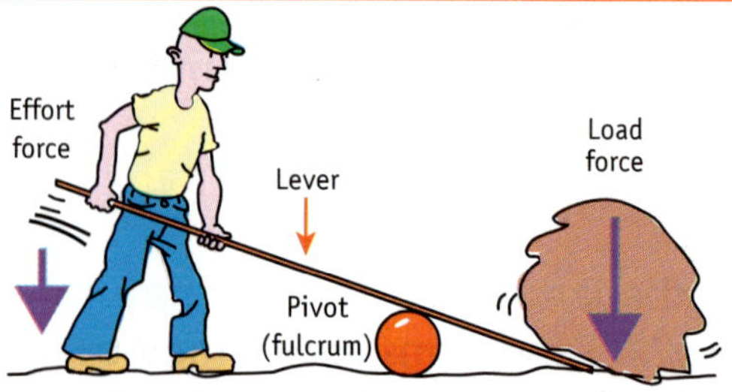

lever types

refer to the placement of pivot, effort and load.

1. The pivot is between the effort *force* and the load force.

Some levers work in pairs (e.g. scissors)

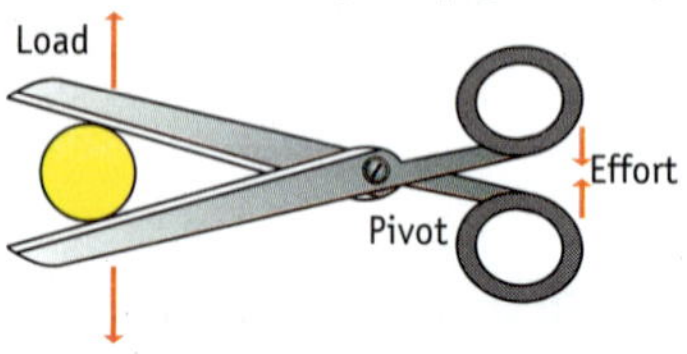

2. The load is between the effort and the pivot. These levers all increase the force.

Wheelbarrow

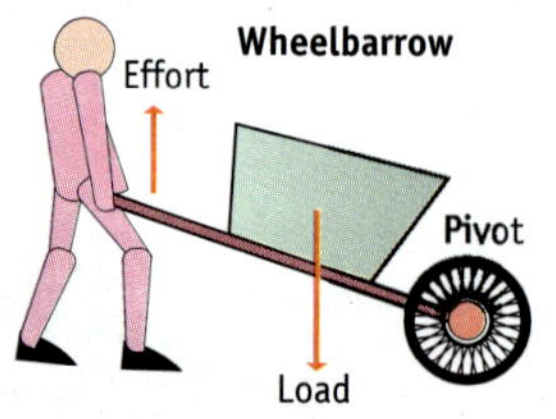

Nutcracker — two levers working together

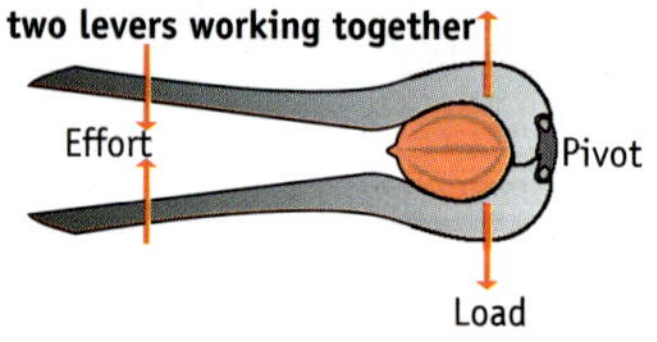

3. The effort is between the pivot and the load. These levers increase movement.

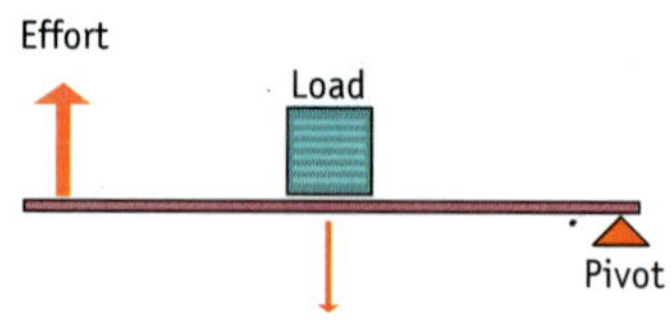

Shovel — used to lift soil

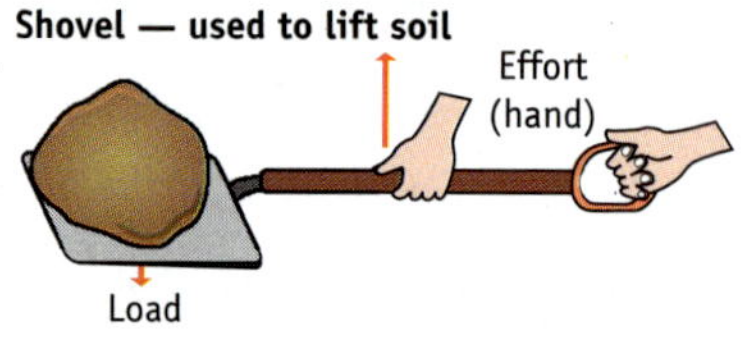

Tweezers — two levers working together

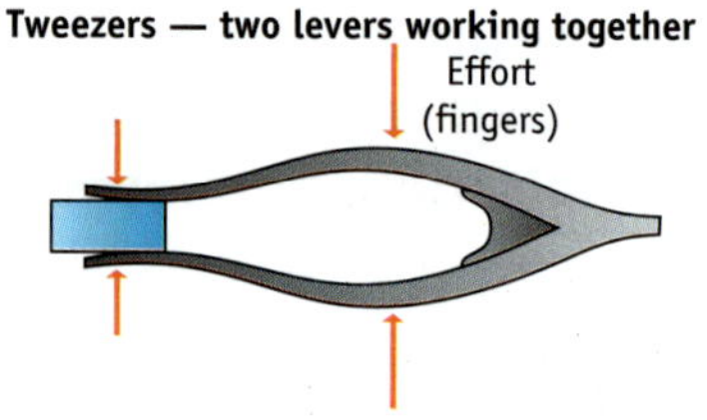

See **levers in the human body**.

levers in the human body

are the lever systems (simple *machines*) formed by the muscles and bones.

Lifting head

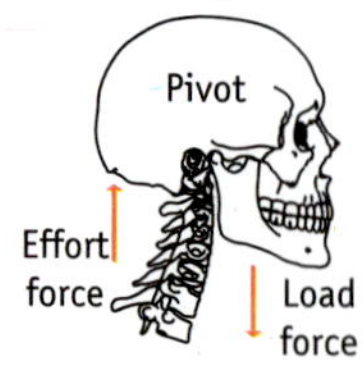

Standing on toe

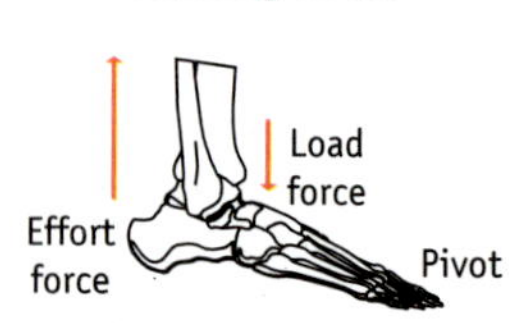

Lifting an object

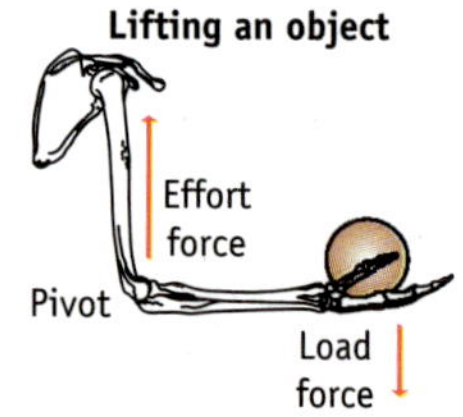

life cycle

is the stages through which *living things* pass as they develop. A young *organism* grows and develops into a mature organism, which reproduces to create more young.
See **life cycle of a flowering plant; life span**.

Life cycle of a moth (e.g. silkworm)

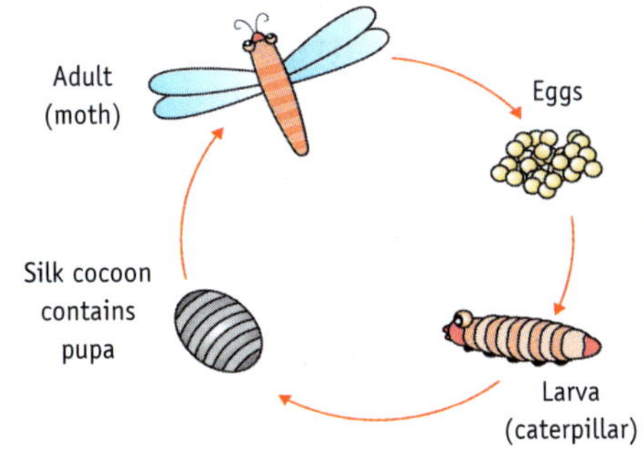

life cycle of a flowering plant

occurs when a *seed* germinates and grows into a mature plant. This plant produces *flower*s where *fertilisation* occurs. As a result, new seeds develop.

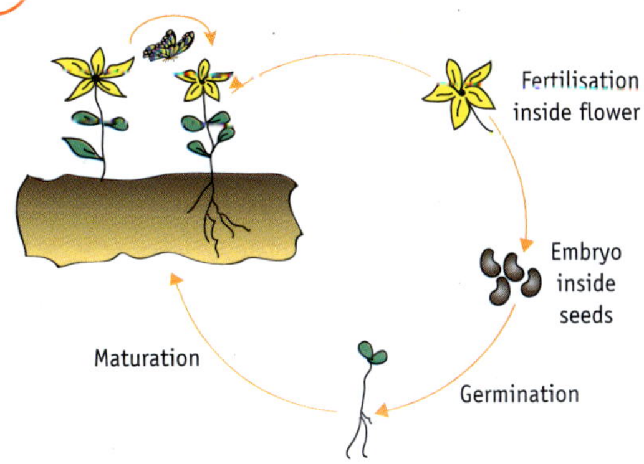

life processes

are activities that occur in *living things*.

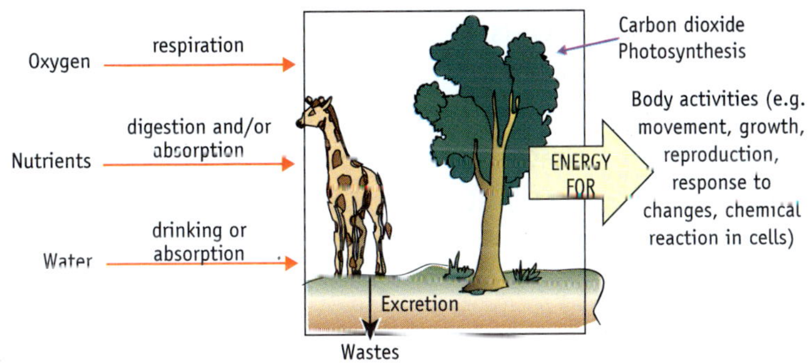

life span

is the length of time an *organism* lives. The life span can be divided into stages where particular changes in the body occur. In humans the life span begins with birth and is followed by childhood, *puberty* and adolescence, during which the person grows and develops. This is followed by adulthood, old age and death. See **life cycle**.

light

is a form of *energy* as it can bring about a change. It is a group of waves, and particles (*matter*) are not transferred. It can travel through a *vacuum*. See **mirrors; lens; light filter**.

light filter

is a screen that allows only some colours to pass through. *Light* is made up of different colours.

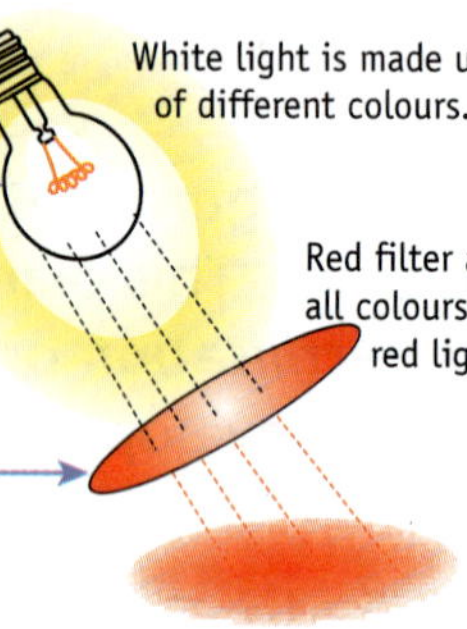

light year

is the distance *light* travels in one *year*. Light travels at 300 000 km/sec. In one year it travels about 9 500 000 000 000 km. Light takes eight minutes to reach *Earth* from the *Sun* (our nearest star). It takes 4.3 years to reach us from the next nearest star.

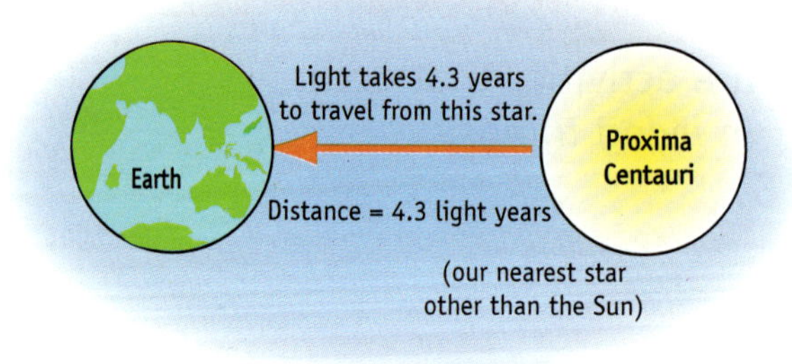

liquid

is one of the three *states of matter*. The particles in a *liquid* can slide over each other so a liquid can flow into a container of any shape. Liquids have a definite *volume*. See **particle theory of matter**.

lithosphere

is the *rock* in the *Earth*'s *crust* and part of the upper *mantle*. It is about 100 km deep. The lithosphere is divided into plates that rub and cause *earthquakes* and *volcano*es, and collide to produce mountains and trenches. See **asthenosphere.** Compare **hydrosphere; atmosphere.**

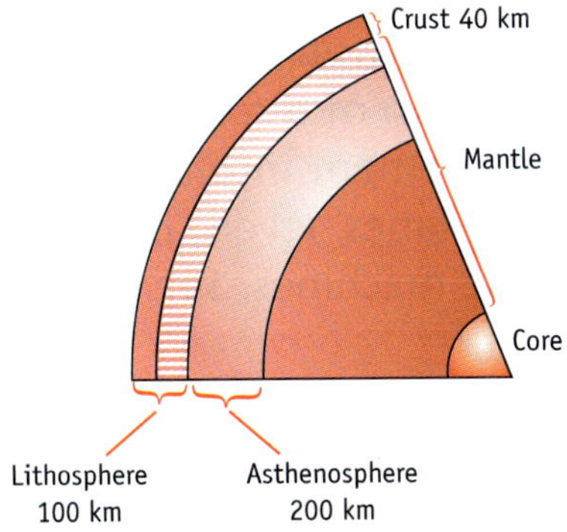

living things

are called *organisms*. They are made of one or more *cell*s and carry out *life processes* (e.g. growing and reproducing). *Animals*, *plant*s, *fung*i and *bacteria* are living things. See **classification of living things**.

loudness (volume of a sound)

is caused by the size of the vibrations. A large vibration causes a loud sound and a small vibration produces a soft sound. Loudness is measured in decibels (dB). See **sound**.

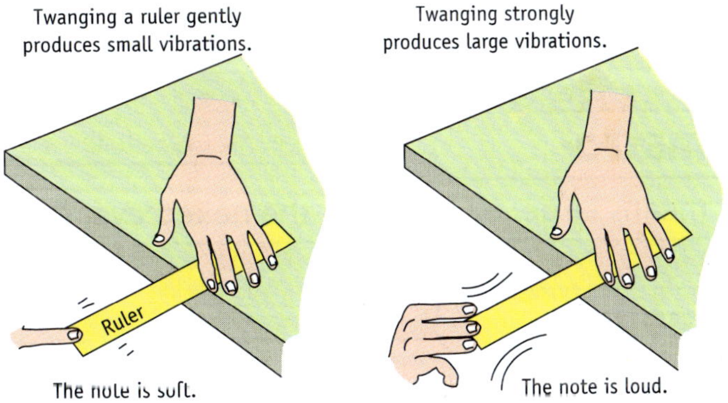

low pressure system

often indicates unsettled *weather*, *rain* and storms. In the southern hemisphere, *air* moves into an area of low *pressure* in a clockwise direction. Compare **high pressure system**. See **wind**.

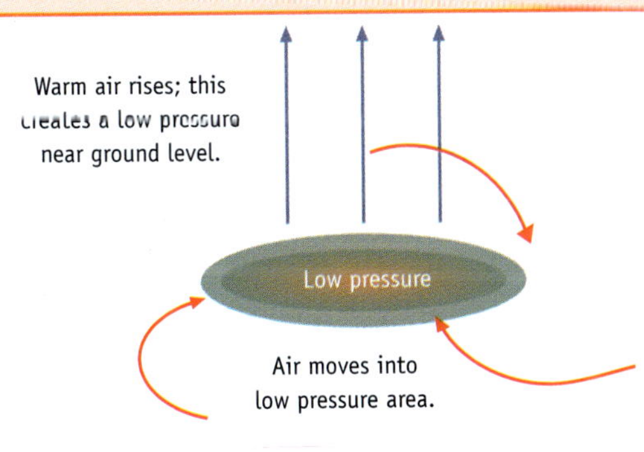

machines

make *work* easier because they can move *energy* to where it is needed, change the direction of a *force* and increase a force. Simple machines can also change a small movement into a large movement and change the *speed* of a force. Inclined planes, *levers*, *pulleys*, *gears*, *ramps* and *hydraulic machines* are examples of simple machines. See **mechanical advantage; efficiency of a machine**.

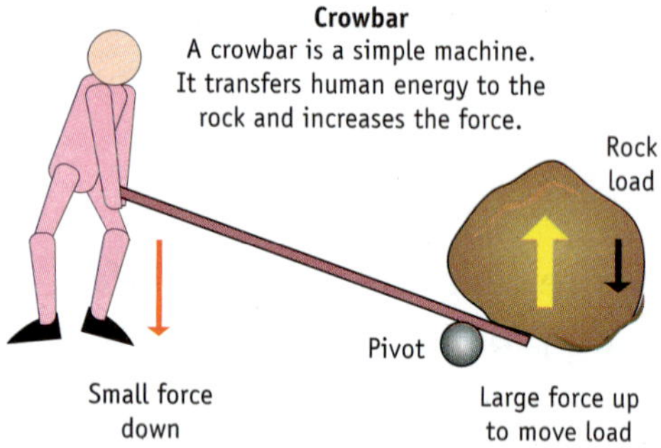

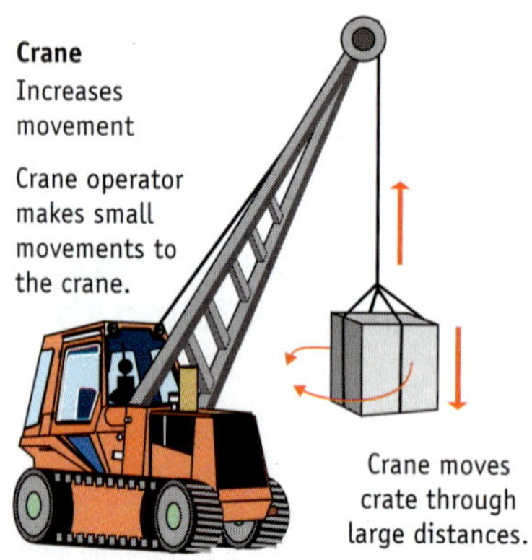

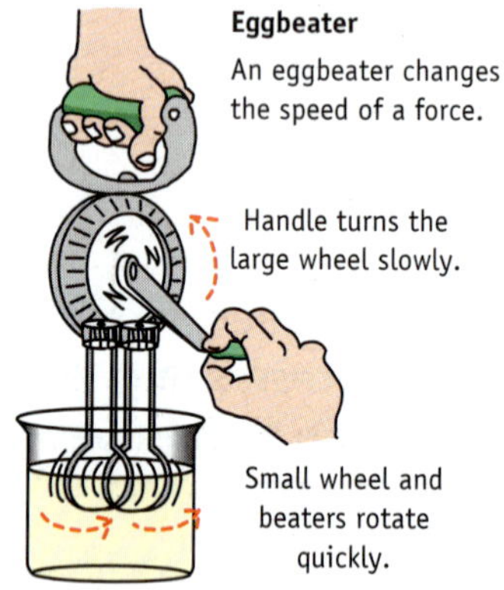

macro-

means large, or visible to the naked eye. Compare **micro-**.

magma

is molten or *liquid rock* found deep inside the *Earth*. Compare **lava**.

magnet

is a piece of *metal* (e.g. iron, steel) that attracts iron and a few other metals. When a *magnet* is hanging freely it will line itself up in a north-south direction.
See **magnetic field; magnetic force**.

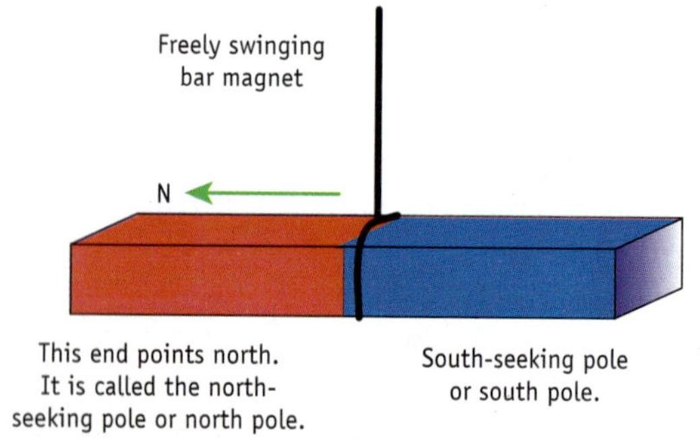

magnet making

is possible by:

- stroking a piece of iron in one direction with one pole of a magnet;
- passing a *current* through a coil with a piece of iron or steel inside the coil.

See **electromagnet**.

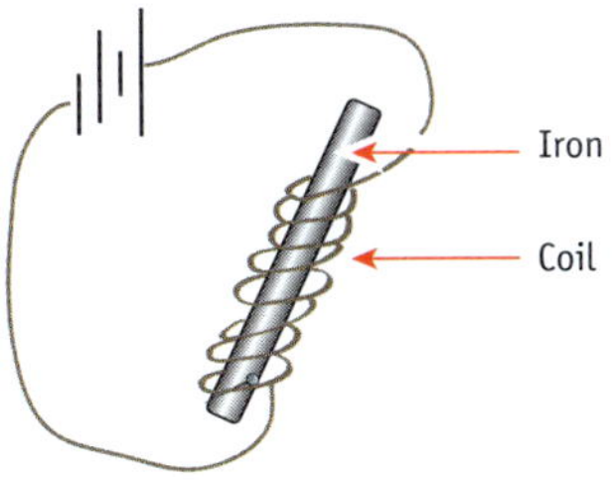

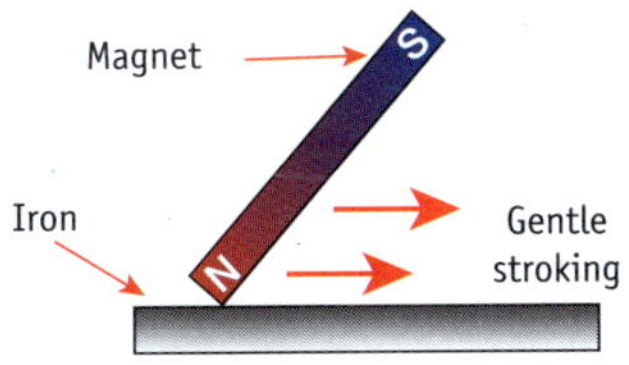

magnet uses

are many. They are used in:

- motors and many household appliances;
- compasses for navigation (a compass is a freely swinging *magnet*);
- *generators* in power stations to make *electricity*;
- *mining*.

See **magnetic separation**.

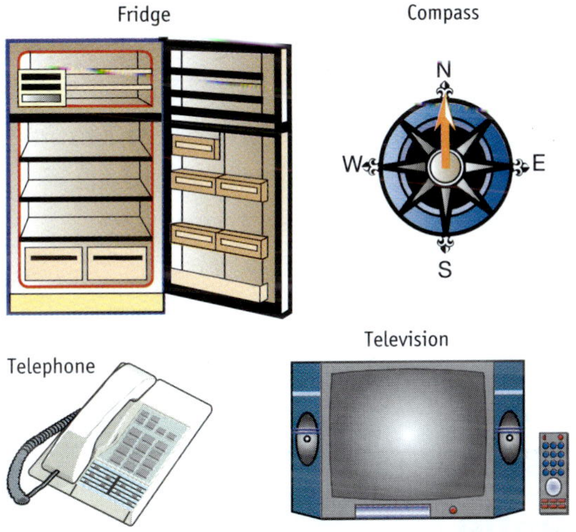

magnetic energy

is the *energy* stored in a *magnet*. It enables the magnet to pick up iron and to attract and repel other magnets. See **magnetic separation; magnetic force**.

magnetic field

exists around a *magnet*. It is the area where *magnetic forces* of attraction or repulsion exist.

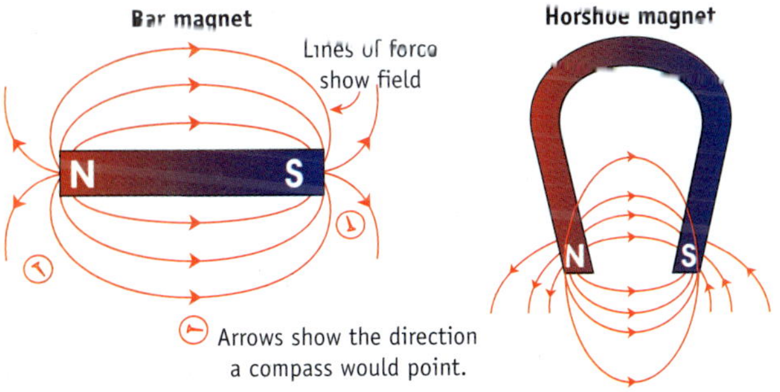

magnetic force

is the *force*s the poles of a *magnet* exert on each other.

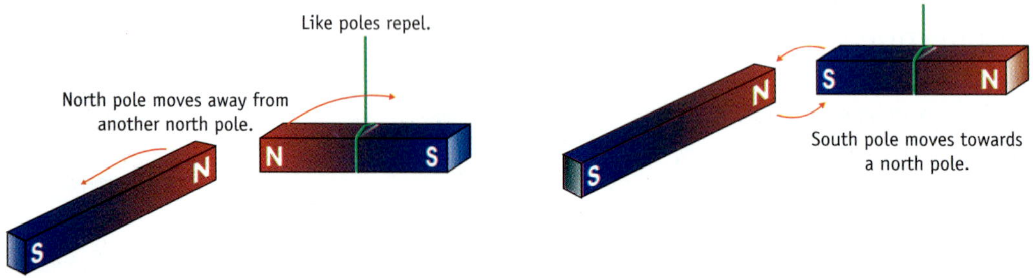

magnetic separation

is when *magnets* are used to separate iron from a *mixture*. Magnetic separation is commonly used in *mining* to separate iron from *rock*s and to remove iron impurities from *ores*.

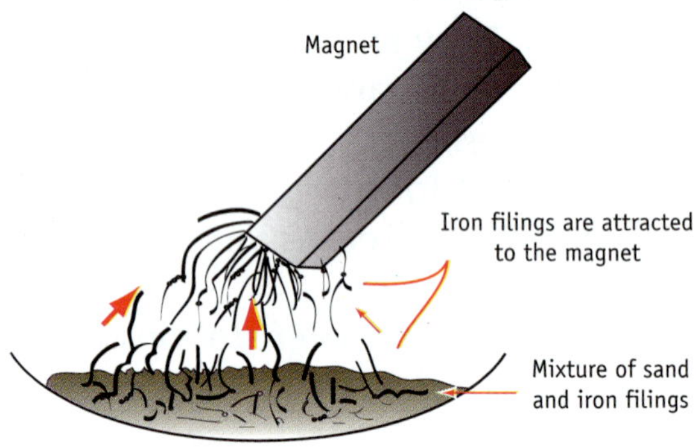

mantle

is the layer of the *Earth* between the *crust* and the *core*. The upper mantle is known as the *asthenosphere*.

manufacture

is the process of making things and *materials* with particular *properties* and for specific purposes. See **plastics; manufactured building materials.**

Cooking tools made of non-stick materials are light and heat-resistant.

This chair is durable and light for carrying.

manufactured building materials

include steel, which is often used to reinforce concrete; aluminium for window frames; building boards such as medium density fibreboard (MDF), plywood and plasterboard (many are made from waste *materials*); bricks, tiles and cement; and glass. See **manufacture**.

marble

is a *metamorphic rock* formed by the recrystallisation of limestone. It often has patterns (veins) and is used for ornamental work and monuments. White marble with fine grain is used for sculpture.

Mars

(the 'red planet') is a rocky *planet* with craters, volcanoes and dried up river beds. The surface is covered with fine red soil, which is blown into the *atmosphere* forming dust storms. Mars has two *moons*.

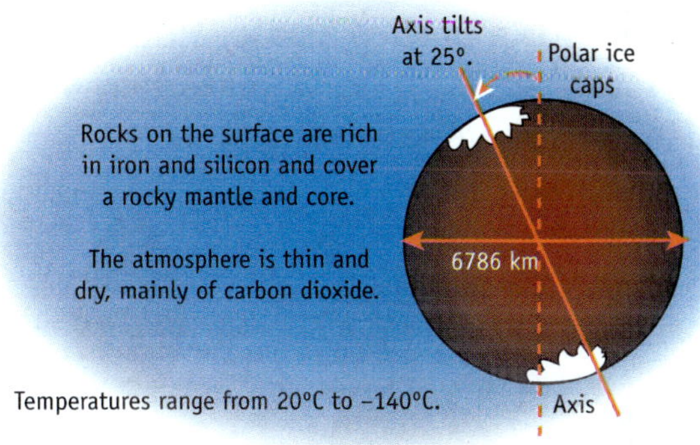

mass

is the amount of *matter* in an object or substance. It is measured in grams (g) and kilograms (kg). Compare **weight**.

material shaping

is a method of strengthening *materials* by shaping them in different ways. Steel beams often have an I-shape; this reduces the *weight* but keeps strength. Folding materials gives them strength. See **material strength**.

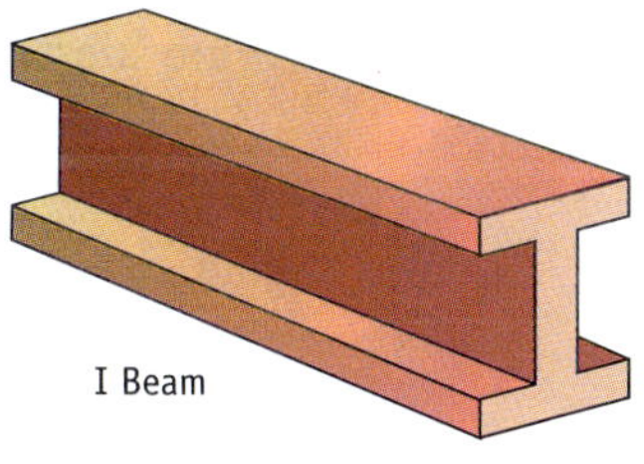

I Beam

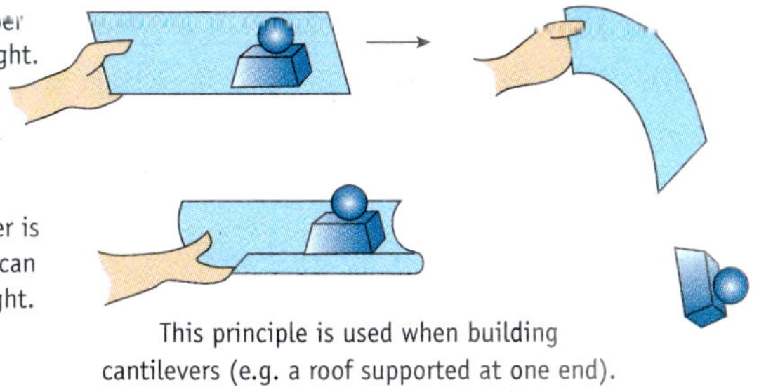

This principle is used when building cantilevers (e.g. a roof supported at one end).

material strength

depends on how the fibres are arranged. *Materials* can be strengthened by lining up fibres in different ways (e.g. plaiting, weaving into fabric or making paper) or by folding. See **material shaping**.

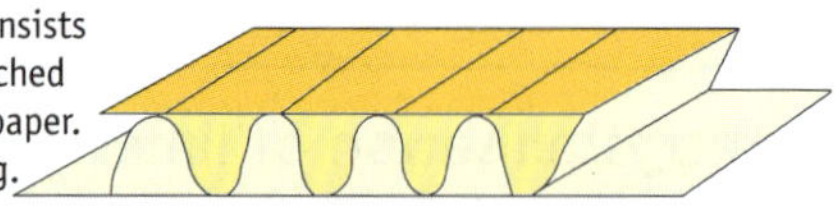

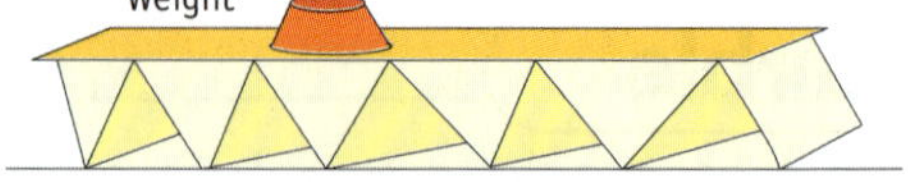

materials

are the substances that make up things. The *properties* of materials make them suitable for a particular use. Humans modify or process materials to make them more useful. See **natural materials; synthetic materials; processed materials; resources; material strength**.

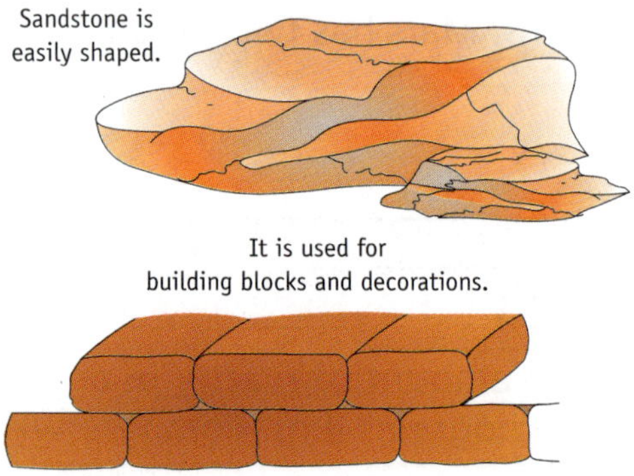

matter

is the material from which things are made. The amount of matter present is the object's *mass*. Matter takes up space and exists as *solid*, *liquid* or *gas*. Solids, liquids and gases occur as *elements*, *mixtures* or *compounds*. See **states of matter; particle theory of matter; forces between particles in matter; expansion of matter; contraction of matter; compression of matter**.

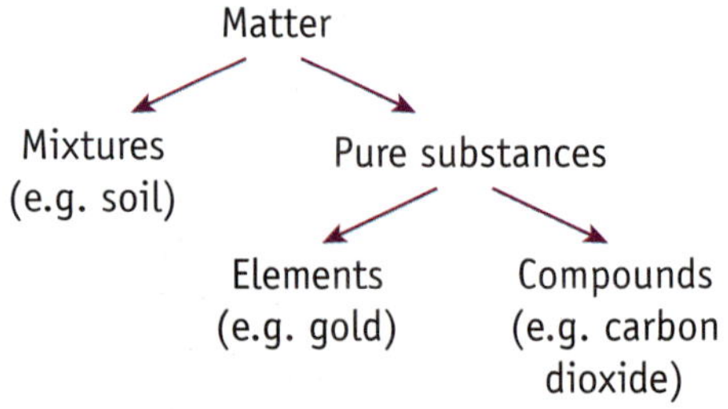

mechanical advantage

of a *machine* compares the load to be lifted with the effort applied to the machine.

$$\text{Mechanical advantage} = \frac{\text{load } \textit{force}}{\text{effort force}}$$

A mechanical advantage larger than 1 indicates the machine is efficient and makes work easier. A mechanical advantage of 1 or less indicates the machine is not very efficient. See **efficiency of a machine**.

megafauna

Are animals that weigh over 500 kg to 1 tonne, or animals that are larger than our domestic animals. An example is the Asian elephant. Most of the megafauna became extinct about 15000 years ago.

melting

is a *change of state* from *solid* to *liquid*. During melting, the solid absorbs heat and its particles begin to move more and slide over each other. See **melting point**.

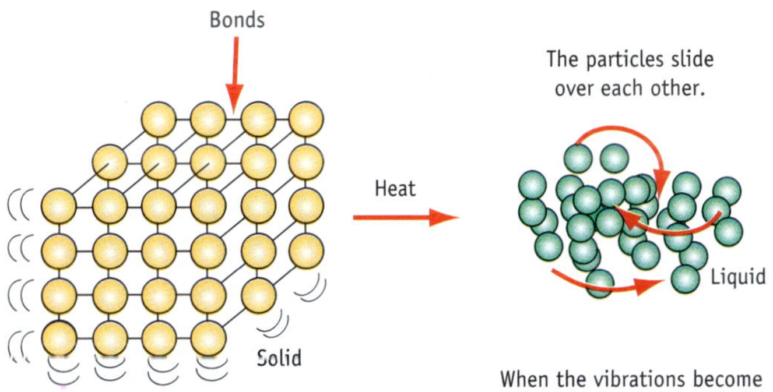

melting point

is the *temperature* at which a *solid* substance turns to *liquid*. A solid with strong *bond*s between its particles (e.g. iron) has a high melting point. A solid with weak bonds has a low melting point (e.g. ice melts at 0°C). Compare **boiling point, freezing point**.

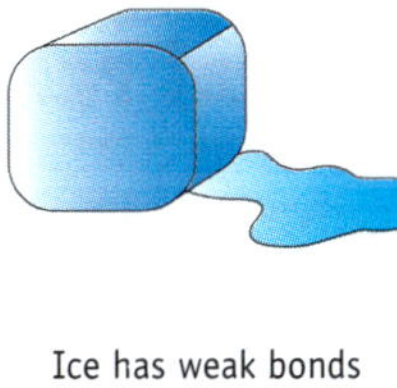

Mercury

is the closest *planet* to the *Sun*. Its surface is similar to that of the Moon. It has craters, plains, cliffs and hills. Mercury is the most iron-rich planet in the *Solar System*. It has no *moons*.

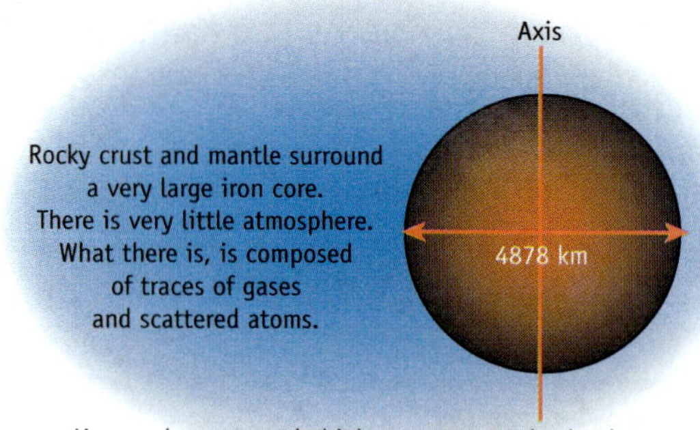

Mercury has extremely high temperatures in the day (350°C) and very low temperatures at night (–170°C).

metal

is strong, shiny (has lustre), can be beaten into a shape and bends without snapping (is malleable). Metals are good *conductor*s of *heat* and *electricity*. See **element symbols; alloys; metal reactivity; metal properties.**

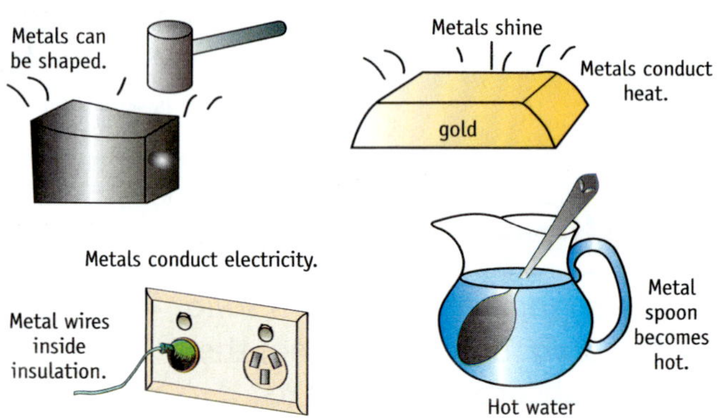

metal properties

are the qualities of the metal. *Metal*s have different *propert*ies and so have different uses. For instance, gold (Au) does not corrode; it is soft and easy to shape. It is not abundant and is used for jewellery, coins and ornaments. Aluminium (Al) is light, easily shaped, not corrosive and abundant. It is used in aircraft and buildings. See **metal reactivity**.

metal reactivity

refers to the different abilities *metal*s have to react with other substances. For example, gold is unreactive, while sodium is very reactive. Unreactive metals may be found in the *Earth*'s *crust* in pure form. Reactive metals exist as *compound*s called *ore*s or *mineral*s.

metamorphic rocks

have formed from other *rocks* that have been altered by *pressure* and/or *heat*. *Earth* movements create massive pressure and heat (from *friction*) and may also alter rocks. See **marble; slate**.

Pressure changes shale to slate.

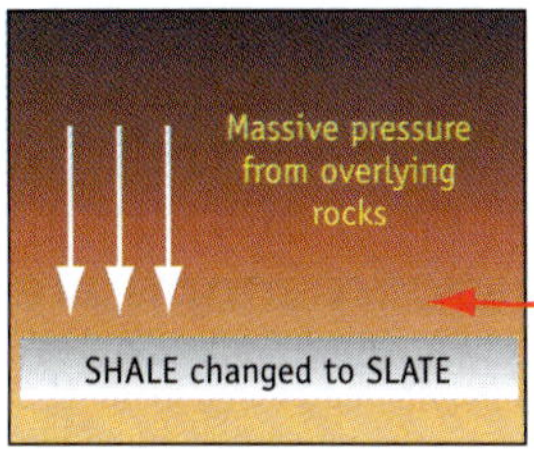

Heat changes limestone to marble.

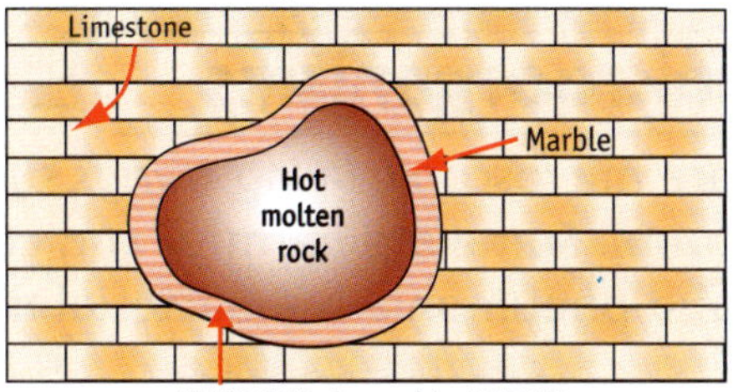

LIMESTONE changed to MARBLE

metamorphosis

is the change of form of an *organism* during its *life cycle*. Some insects change from *larva* to *pupa* to adult. Adult frogs develop from tadpoles. See **life cycle**.

meteorites

are lumps of *rock* or *metal* that come from *Space* and hit the *Earth*. Compare **meteors**.

meteorology

is the study of the *atmosphere* in relation to *weather* and *climate*.

meteors

or 'shooting stars' are small fragments of *matter* attracted to the *Earth* from *space*. As they enter the *atmosphere*, *heat* from *friction* causes them to burn. Compare **meteorite**.

micro-

means small and often refers to structures that cannot be seen with the naked eye (e.g. *microorganism*).

microorganisms

are microscopic *living things* such as *unicellular organisms*, *bacteria* and some *fungi*. Some are harmful and cause disease. For example, tinea is a fungal disease and pneumonia is a bacterial disease. Many microorganisms are useful.

Useful microorganisms

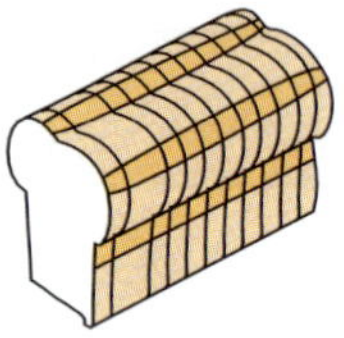

The fungus, yeast, is used to bake bread.

Bacteria in the gut of herbivores digest plant food.

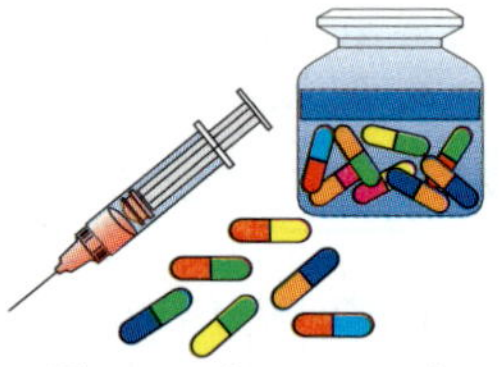

Microorganisms are used to make vaccines and antibiotics which help control disease.

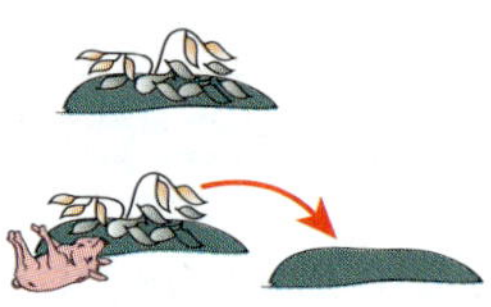

Microorganisms decompose dead organisms and recycle nutrients.

Milky Way

is the band of *stars* and hazy *light* in the night sky. It is part of our *galaxy*.

mineral

is a *compound* found in the *Earth* in *rocks*. Common minerals are quartz, feldspar and calcite. See **mineral properties; mineral uses; mineral nutrients; rock minerals; ore minerals**.

mineral nutrients

are needed for tissue growth and body functioning (e.g. *animals* need iron to make blood and *plants* need nitrates to make *protein*). Plants obtain *minerals* from the *soil* and animals obtain them from their *food* (i.e. plants or animals).

Calcium

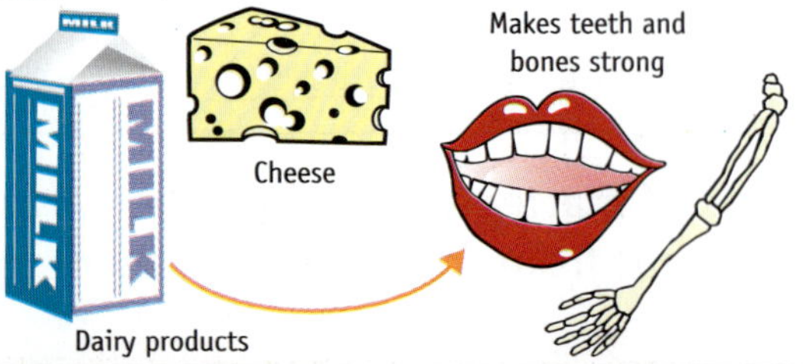

Magnesium

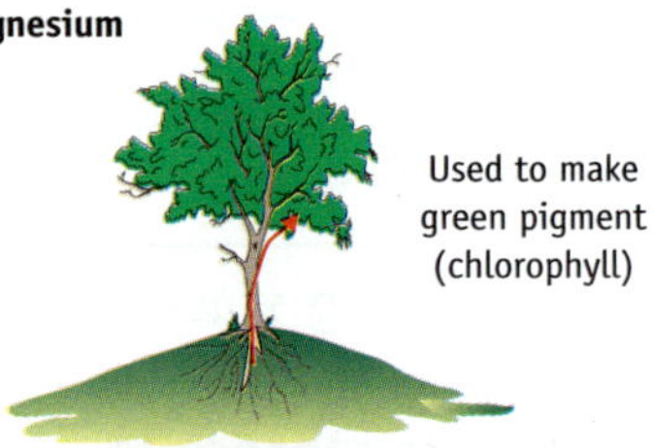

mineral properties

allow *minerals* to be identified. The *properties* are their distinct *crystal* shape (which may be microscopic), colour, colour of their streak (powder), hardness, lustre or sheen, and cleavage (the way they split).

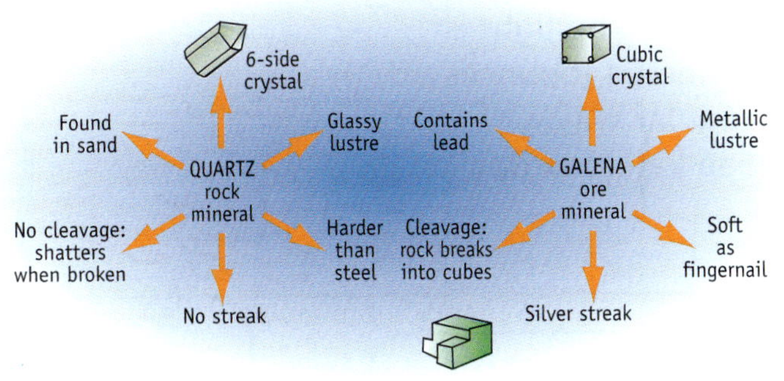

mineral uses

are related to their *properties*.

Mineral	Property	Use
Quartz	hard	sandpaper
Ruby	attractive	jewellery
Haematite	coloured	pigment and dye
Mica and asbestos	*heat*-resistant	*insulators*

mining

is used to extract *minerals* and *ores* and other useful substances, such as *coal*, from the *Earth*. Mining destroys small areas of natural land and *ecosystems*. See **non-renewable resources**.

mirrors

are reflecting surfaces. They may be flat (plane) or curved. See **mirrors, reflection; mirrors, uses**.

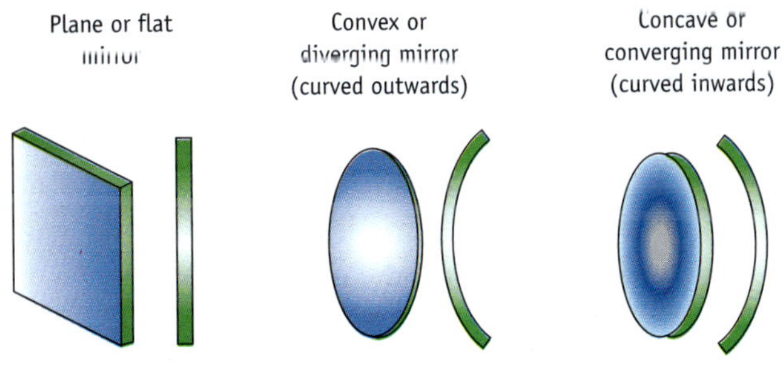

mirrors, reflection

occurs when *light* hits a *mirror*. The light is reflected (bounces off) in a regular way. See **kaleidoscope**.

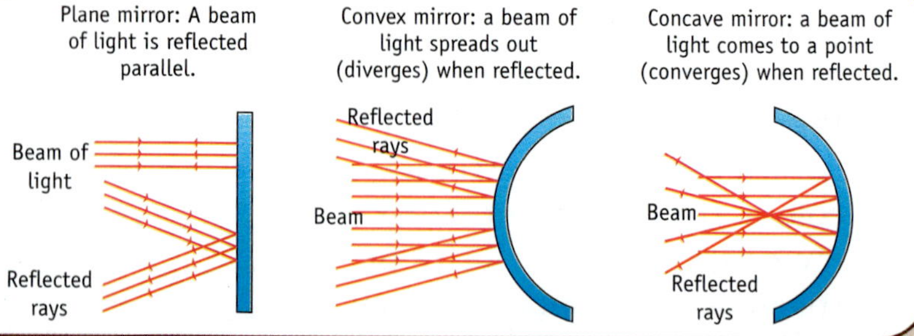

mirrors, uses

depend on the shape of the mirror. With plane (flat) mirrors the image is undistorted, reversed sideways and is the same size as the object (e.g. bathroom mirror). With convex mirrors the image is smaller than the object and there is a wide field of view. They are sometimes used for rear vision mirrors on cars. With concave mirrors the image is larger than the object so these mirrors are often used for shaving or make-up. They are also used in *telescopes*, reflectors for solar furnaces and in search lights and torches. See **mirrors; mirrors, reflection**.

mixture

is made of two or more substances mixed or blended together without a chemical change (e.g. sea water, *soil*, *air*). *Solutions*, *suspensions*, *colloids* and *emulsions* are mixtures. The substances in a mixture may be separated because of their different *properties*. Compare **element**; **compound**. See **separation methods**.

Ingredients are mixed to make a cake.

Mixture of flour, sugar and butter.

molecule

is a group of *atoms* held together by forces or bonds. It is the unit that makes up a compound (e.g. sugar is a *compound* made up of sugar molecules). The atoms in the molecule may be the same or different.

A hydrogen molecule H_2 is made up of 2 hydrogen atoms

A water molecule H_2O is made up of 2 hydrogen atoms and 1 oxygen atom

monoculture

is the growing of one type of crop (e.g. wheat) over a large area. It creates an unbalanced *ecosystem*.

moon

is a large natural body or *satellite orbit*ing a *planet*. Our Moon is about one-quarter the size of the *Earth*. It has no atmosphere and no water, is rocky and covered with soil and the surface is made up of craters, plains ('seas'), mountains and valleys. The side facing the *Sun* reflects the Sun's *light* rays. This causes it to glow.

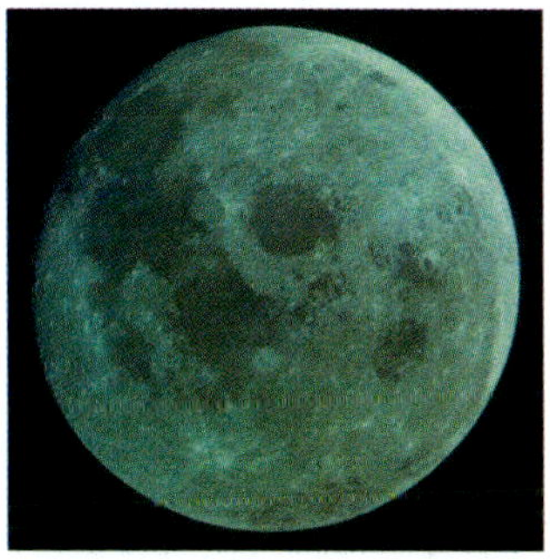

Earth's Moon

moulting

is the shedding of skin (reptiles) or exoskeleton (insects) to enable the *animal* to grow, or the shedding of feathers (birds) or hair (mammals) to allow new growth. Birds and mammals moult when they shed feathers or fur (hair).

movement energy

See **kinetic energy**.

multicellular organisms

are made up of many different *cells*, which are usually grouped into *tissues* and *organs*. Compare **unicellular organisms**.

natural disaster

is a severe natural event that destroys the natural *environment*, people and property. See **drought; bushfires and floods; cyclones; earthquakes; volcano; tsunami; El Niño.**

natural environment

See **habitat**; **ecosystem**.

natural materials

come from *living things*, the *Earth*, *air* or water (e.g. *food*, *ores*, *oxygen*, wood). See **materials**; **resources**.

Many natural *materials* are modified to make them more useful (e.g. raw wool and cotton are converted to useful fibres). See **processed materials**.

Humans use natural *materials* to make new materials (e.g. *plastics*, *detergents*, fibreglass). See **synthetic materials**.

nebulae

are large clouds of dust and glowing *gas* in *space*. The Crab Nebula is a huge cloud of expanding gas formed by an exploding *star* (supernova). The Orion Nebula glows because of newly formed stars embedded in the gas.

Neptune

is a very distant *planet* in our *Solar System*. It is made of *gas* and has distinctive cloud features with feint belts and zones. It has a large spot, called the Great Dark Spot, and two smaller ones caused by storms. There are eight known *moons* orbiting Neptune.

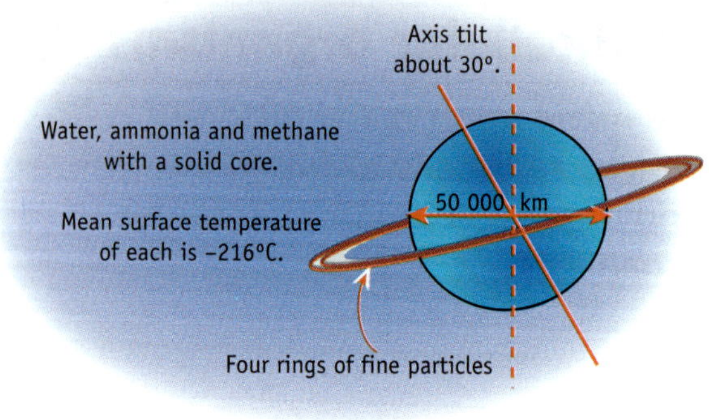

nocturnal

describes *animals* that move around and obtain *food* mainly at night (e.g. owls, possums). The darkness provides protection and they are able to avoid the *heat* of the day. Compare **diurnal**.

non-metals

are generally light, do not have lustre (sheen) and are not malleable (cannot be beaten into a shape). They do not conduct *electricity* or *heat*. Carbon and *oxygen* are non-metals. About one-sixth of the *elements* are non-metals. Compare **metals**. See **elements; element symbols**.

non-renewable resources

are substances that cannot be replaced once they are used (e.g. *fossil fuels*, *minerals*, *rock*, *metals*). Compare **renewable resources**. See **energy sources, non-renewable; conservation of non-renewable resources**.

nuclear energy

is released from radioactive substances, such as *uranium*. In nuclear power plants, nuclear energy is used to make *electricity* and products important for medical *science*.

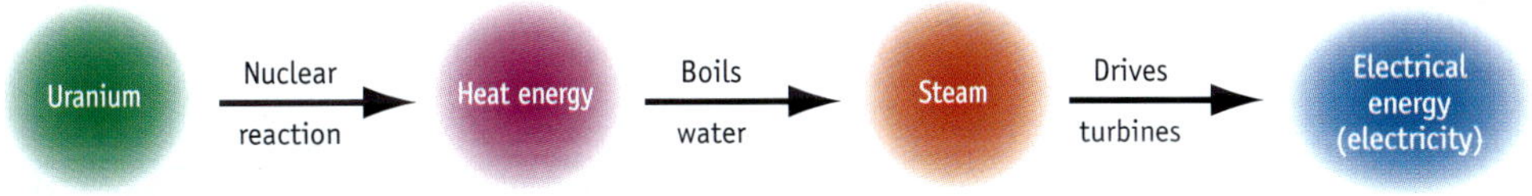

N

nutrients

are the *food* substances an *organism* requires to function correctly and to maintain health. There are five main groups: *proteins*, *carbohydrates* (sugars), *fats*, *vitamins* and *minerals*. See **organic compounds; inorganic nutrients**.

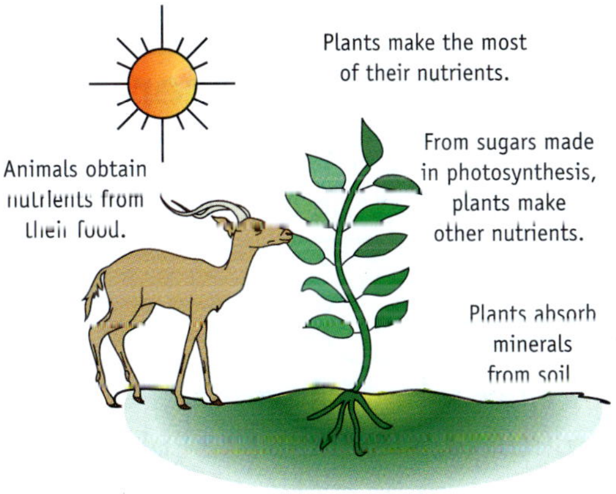

nutrition

is the process by which organisms absorb substances (*nutrients*) needed for maintenance, growth and repair of *tissues* and functioning of the body.

O

ocean currents

ocean currents

are the movements of the ocean water caused by *winds* and different water densities. The *density* of the water changes as it heats or cools and as the salt content changes. See **global ocean currents**.

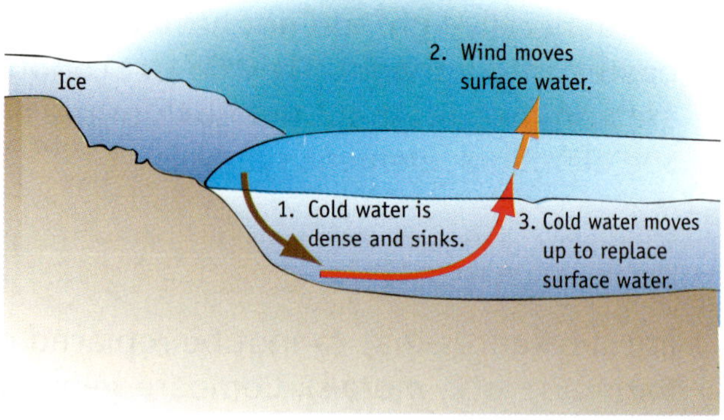

ocean depth

affects the conditions present. *Animals* living in the depths of the ocean have *adaptations* that enable them to survive without *light* or *plants*, in low *temperatures* and under high *pressures*.

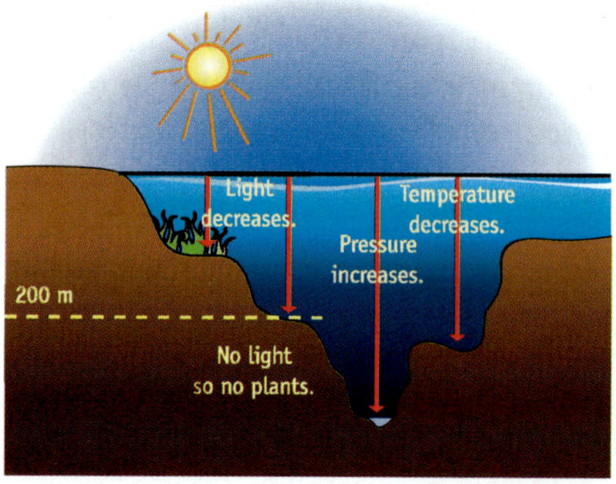

oceans and weather

are connected. The ocean waters cool or warm the *air* and add moisture to it. They influence the *weather* and *climate*. See **global ocean currents; El Niño**.

opaque

describes something *light* cannot pass through (e.g. wood). The light bounces off (is *reflected*) or is absorbed. When light hits an opaque object, a *shadow* may be formed. Compare **transparent; translucent**.

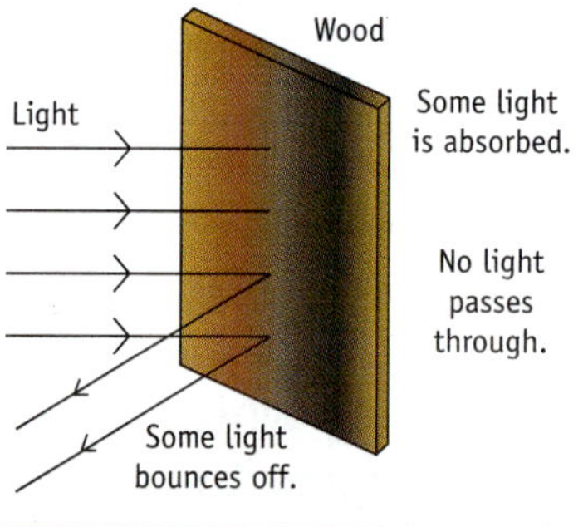

orbit

is the regular path of one object around another (e.g. *planets* and *satellites* are in orbit). The planets move around the *Sun* in elongated orbits (ellipses), in the same direction and almost the same plane (except *Pluto*). See **year**.

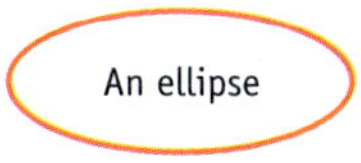

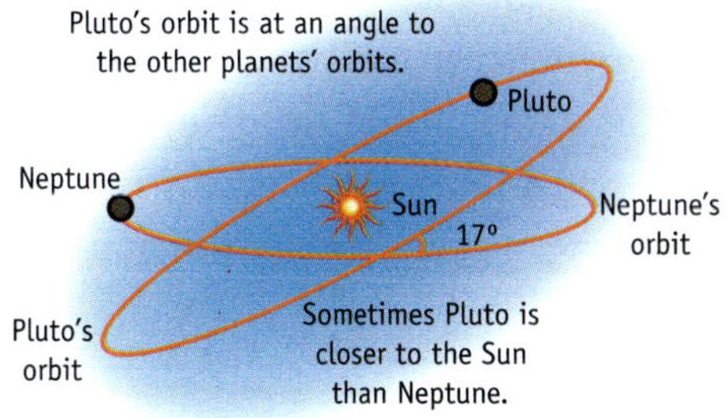

ore

is a concentrated mass of useful *minerals* mixed with unwanted *materials* and found in the *Earth's crust*. Haematite is an ore containing iron. Bauxite contains aluminium. See **mining; refining**.

ore minerals

are *minerals* found in *ores*. They are *metals* and other substances useful to humans. For example, aluminium is found in bauxite and iron in haematite. Titanium is obtained from the ore rutile and copper from chalcopyrite.

organ

is part of a living body made up of one or more *tissues*. Different organs work together as *organ systems*. For example, the *digestive system* is made up of organs such as the stomach, liver and intestines.

organ systems

in humans work properly when the person has a healthy lifestyle (a well-balanced diet, exercise and rest). *Organ* systems in *plants* include *stems*, *roots* and *leaves*. See **plant structure**.

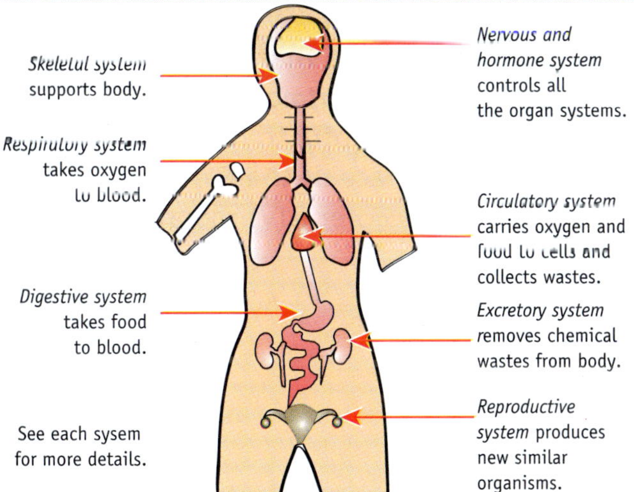

organic compounds

contain carbon and hydrogen and many have a long chain structure. They make up living *tissues* and are found in *materials* from *living things* (e.g. wood, petrol, cotton, wool, leather). *Proteins*, *carbohydrates* (sugars), *fats* and *vitamins* are organic *nutrient*s. Compare **inorganic nutrients**.

organism

is a *living thing*. See **multicellular organisms; unicellular organisms**.

ovum (plural: ova)

See **egg**.

oxygen (O_2)

is a *gas* found in the *atmosphere*. It is used by *plant*s and *animal*s for *respiration*, and to burn substances.

ozone (O_3)

is a form of *oxygen* that exists in a layer in the upper *atmosphere*. This ozone layer is helpful to life as it shields the surface of the *Earth* from excessive *ultraviolet* (UV) radiation from the *Sun*.

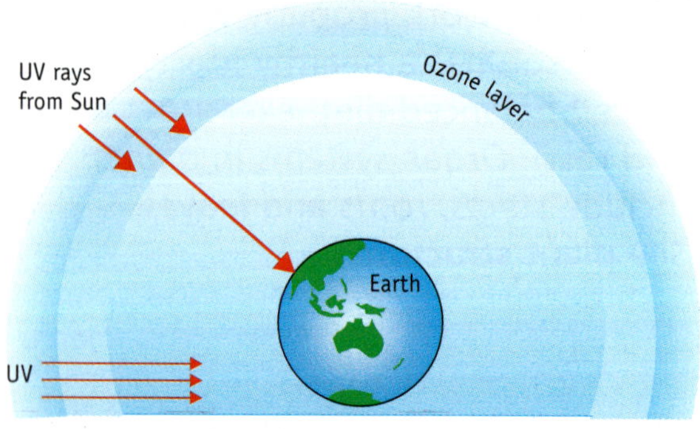

Some pollutants (CFCs) reduce the ozone layer and allow more UV to reach the Earth.

parasite

See **parasite**ism.

parasitism

occurs when one *organism* lives on or in another and feeds from it.

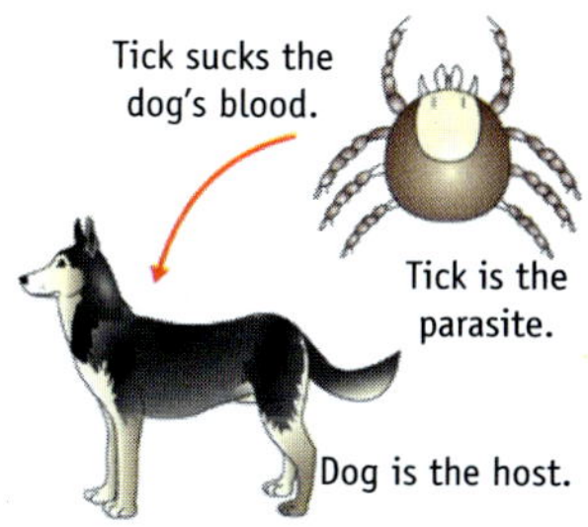

parental care

is the care mammals, birds and some reptiles give their young after birth so they have a high survival rate. Fish and amphibians do not care for their young and so the young have a high death rate. These creatures produce large numbers of offspring to ensure some survive.

parenting

See **parental care**.

particle theory of matter

explains the behaviour of substances. This theory describes *matter* as small, moving particles (*atoms* or *molecules*). Heating matter makes the particles move more rapidly and cooling it slows their movement.

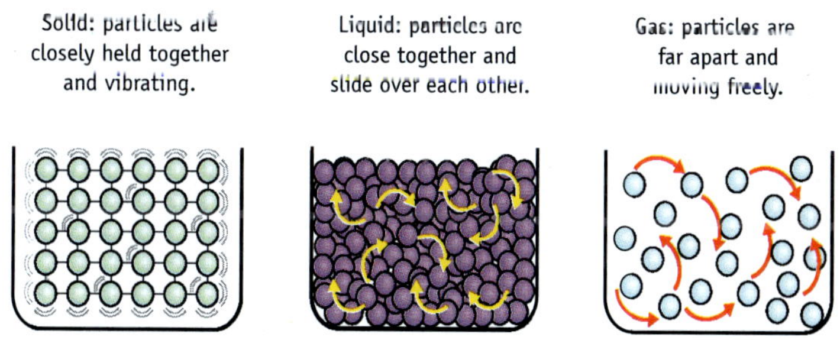

petroleum

is a *fossil fuel*. It is refined to obtain petrol and many other useful substances. See **fossil fuel uses**.

Petroleum formation

pH

is a measure of acidity. It ranges from 1 to 14. Acid substances have a pH below 7, neutral substances have a pH of 7 and alkaline substances have a pH above 7. Universal indicator shows the level of acidity.

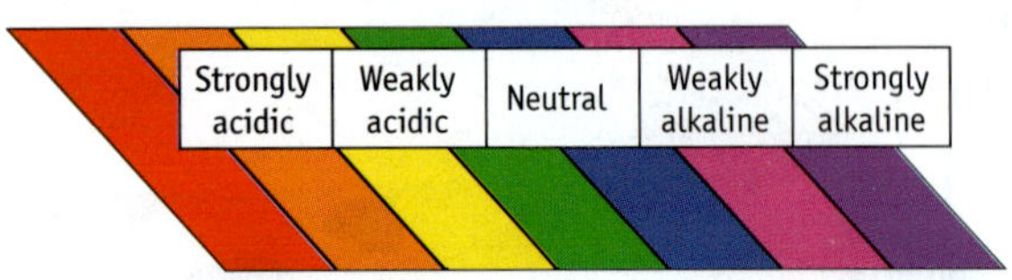

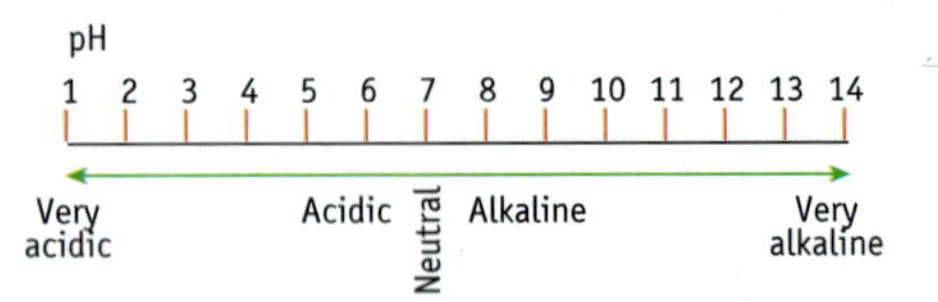

phases of the Moon

refers to its different shapes (e.g. Full Moon, New Moon). The side of the *Moon* facing the *Sun* reflects light (i.e. is lit up). As the Moon revolves around the *Earth*, we can see different portions of the Moon reflecting light to the Earth.

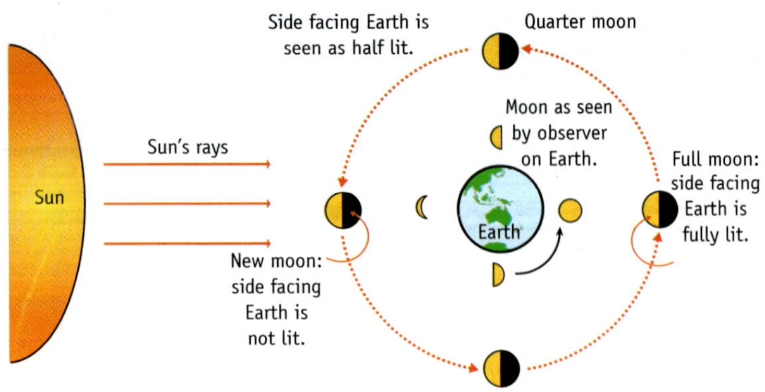

phloem tissue

carries the *food* made in the leaves of *plant*s to *cell*s and *organ*s for use or storage (e.g. sugars are stored in *fruit*s). See **transport systems in plants**.

photosynthesis

is the process *plants* use to make their own *food*. In this process, the plants use sunlight to convert *carbon dioxide* and water to sugars (food). At the same time, *oxygen* and water are formed. See **chloroplast**.

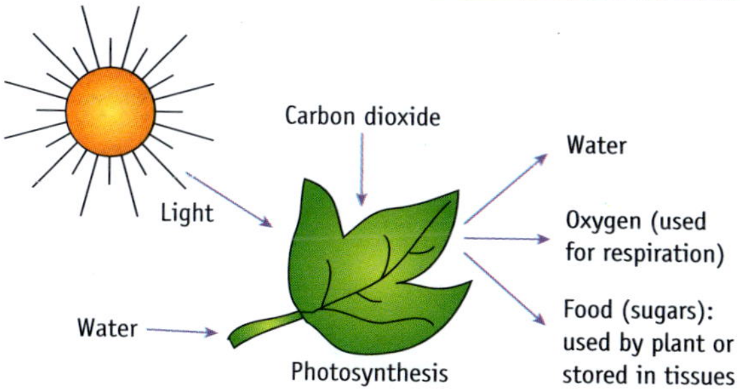

physical change

occurs when *matter* changes but no new chemical is formed, that is, there is a *change of state* (e.g. wood chopped into pieces, sugar *dissolved* in water). Generally it can be readily reversed. Compare **chemical reaction or change**.

physical (mechanical) weathering

is caused by several processes.

1. Abrasion by particles carried by *wind*, water or *glaciers* wear down the rock. See **corrasion**.
2. *Temperature* changes make *rocks* expand and contract, weakening them so that they crumble. Water freezing in cracks in rocks also splits them.
3. The roots of *plants* grow into cracks.

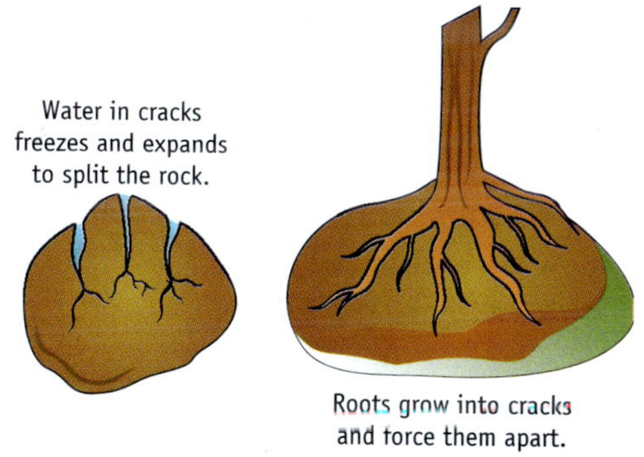

pitch of a sound

is its highness or lowness.
A rapid vibration produces a high note and a slow vibration produces a low note. The number of vibrations in a second is called the frequency.
See **sound energy**.

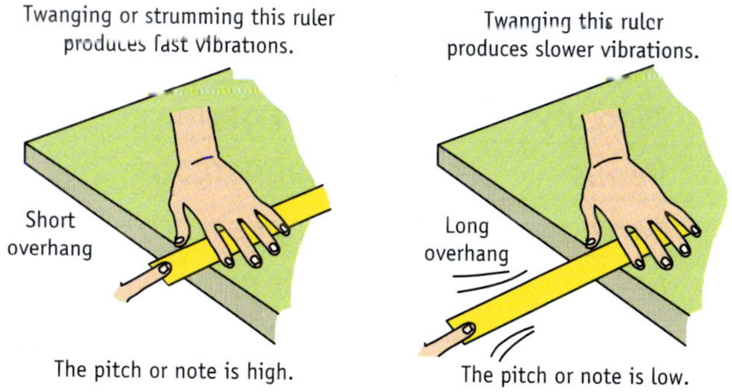

planet

is a body revolving around a *star*. In the *Solar System* nine planets *orbit* the *Sun*. The planets, in order from the Sun, are *Mercury*, *Venus*, *Earth*, *Mars*, *Jupiter*, *Saturn*, *Uranus*, *Neptune* and *Pluto*. Each rotates on its axis and most have *moons*. See **planet sizes; planet distances from the Sun**.

planet distances from the Sun

are immense. The distance between *Pluto* and the *Sun* is 5900 million km. If we call this distance 100 mm (10 cm), then the distance of each *planet* from the Sun is:

Planet	Distance	Planet	Distance
Mercury	1 mm	*Saturn*	24 mm
Venus	2 mm	*Uranus*	50 mm
Earth	2.5 mm	*Neptune*	77 mm
Mars	4 mm	*Pluto*	100 mm
Jupiter	13 mm		

planet sizes

are also very large. The diameter of the *Sun* is 1 400 000 km. If we imagine the Sun is the size of a large rockmelon, then the *planet*s would be about these sizes.

Relative sizes of planets

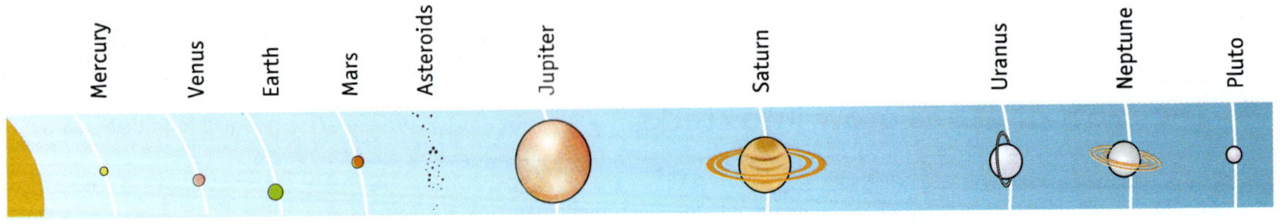

Planet	Size	Planet	Size
Mercury	salt grain	*Saturn*	large grape
Venus	peppercorn	*Uranus*	pea
Earth	peppercorn	*Neptune*	small pea
Mars	salt grain	*Pluto*	salt grain
Jupiter	walnut		

plant

is an *organism* that makes its own *food* by the process of *photosynthesis*. Its *cells* have a cell wall and many contain *chloroplasts*. The main groups of plants are algae (mainly water weed), mosses and liverworts, ferns, cone plants or conifers, and *flowering plants*.

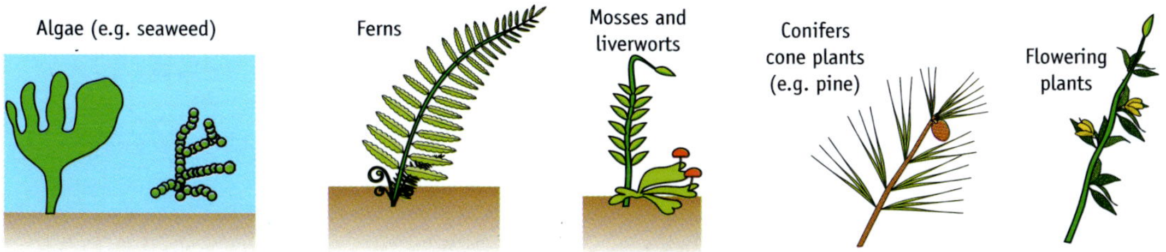

plant structure (flowering plant)

consists of *leaves*, *stems*, *roots* and *flowers*. These *plant organs* provide the plant with *nutrition* and enable it to reproduce and function (*work*).

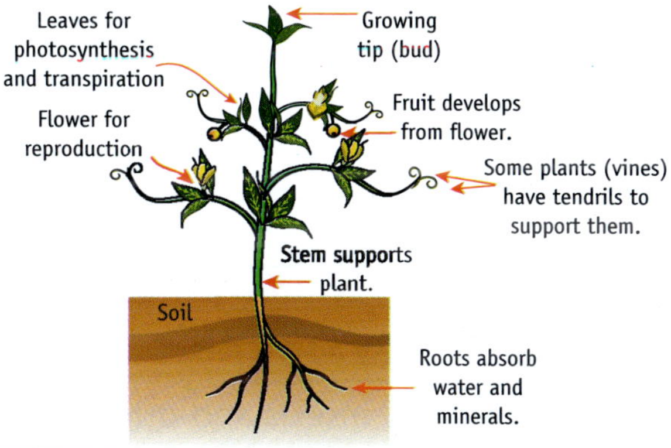

plastics

are made by humans (*synthetic materials*) and most are made from *petroleum*. They are organic and generally have very large *molecules*. There are many different types and they have many uses.

Plastic	Properties	Use
PVC	hard-wearing, strong	floor and upholstery coverings, guttering
Polystyrene	can be made into light foam	packaging, insulation
Nylon	light, strong; can be made into threads	ropes, clothing, fishing lines
PTFE	*heat*-resistant; resists chemicals	non-stick lining on frying pans
Polythene (polyethylene)	tough, flexible; resists chemicals	food wrapping, shopping bags

Pluto

is the furthest *planet* from the *Sun* in our *Solar System*. It is tiny and too far away from *Earth* to see surface details. Pluto has a very large, close *moon* and together they *orbit* the Sun on a plane different from that of the other planets.

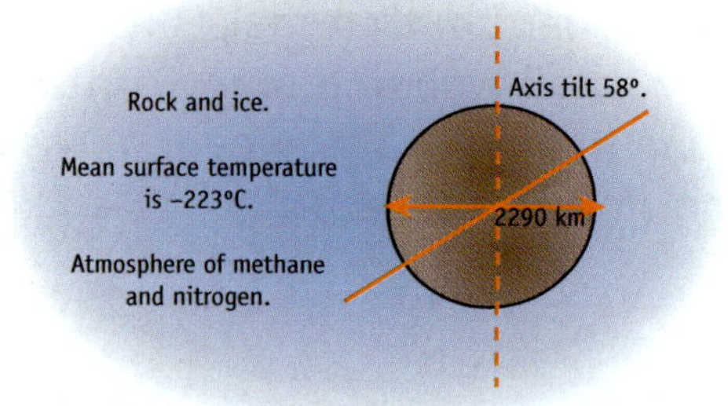

pollination

is the transfer of pollen, which contains the male sex *cell*, to the female part of the *flower* (the stigma). Pollination enables *plant*s to reproduce.
See **pollination by animals; pollination by wind**.

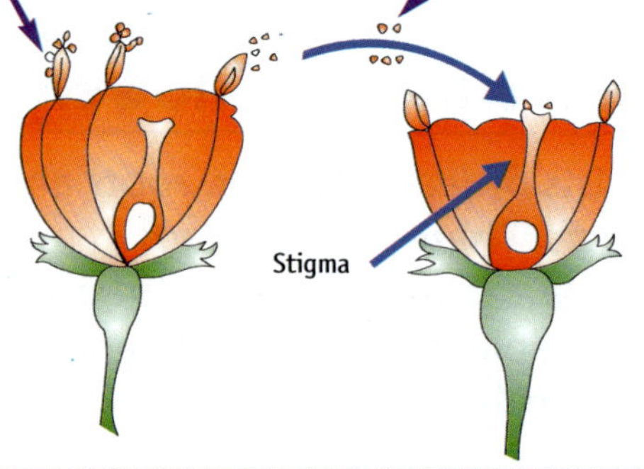

pollination by animals

occurs when birds, insects and other *animals* transfer the pollen while they are foraging for nectar. The *flower*s of plants pollinated by animals are often brightly coloured and/or have strong perfume to attract *pollinators*. Compare **pollination by wind**. See **pollination**.

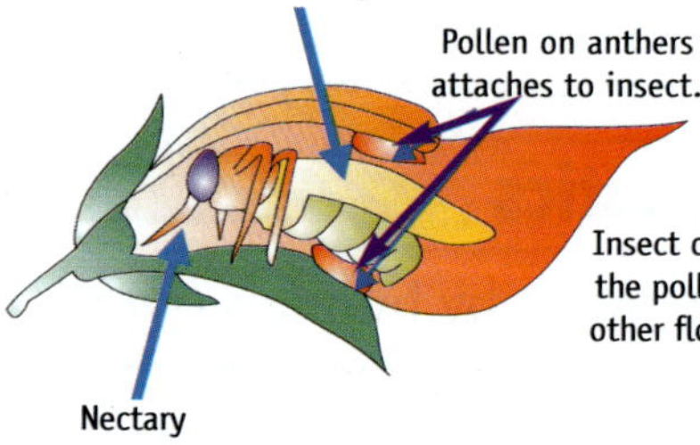

pollination by wind

is used by some *plant*s (e.g. grasses). Their *flower*s tend to be small and plain but well exposed to catch air-borne pollen. Compare **pollination by animals**. See **pollination**.

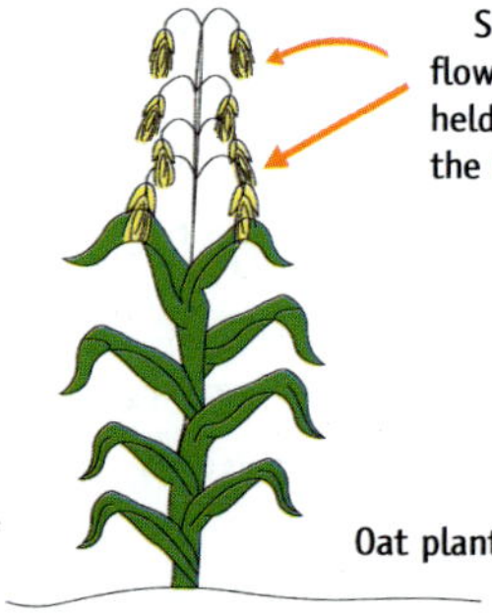

pollinators

See **pollination by animals**.

pollution

is the spreading of harmful substances (e.g. fumes from cars, smoke from factories and refineries, garbage, sewage, pesticides and fertilisers) into the *environment* and the use of *materials* that are not *biodegradable* (e.g. many *plastics*). See **global warming; waste disposal.**

population

is a group of similar *organisms* living in an area (e.g. the rabbit population in a paddock, koala population in a national park). *Environmental changes* may cause the numbers in the population to change. See **population numbers**.

population numbers

may change if new *species* are introduced into the area. These new *organisms* may *prey* on or compete with the original *population* so that their numbers drop, or they may provide *food* for the original population so that population numbers increase. See **predation**.

porous

means full of small holes or openings (e.g. a sponge). Sandy *soils* are porous and water drains through them rapidly. Less porous soils (e.g. clay soils) hold or retain more water. See **soils, water retention**.

potential energy

is stored *energy*. A rock on a cliff, a wound-up spring, a stretched elastic band, chemicals in *batteries*, *fuel* and firecrackers, all have potential energy. See **energy transformation**.

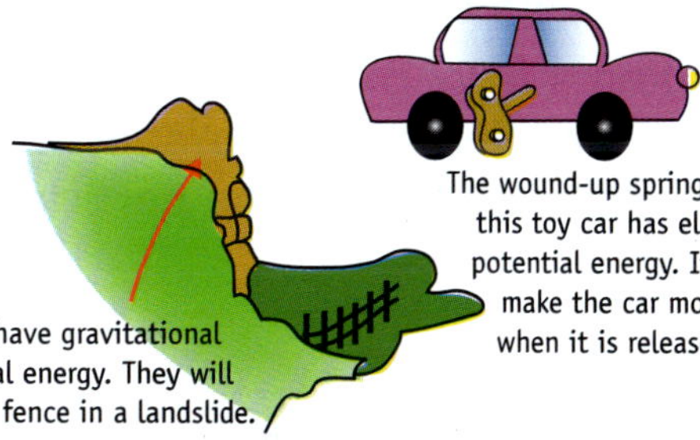

Rocks have gravitational potential energy. They will crush the fence in a landslide.

The wound-up spring inside this toy car has elastic potential energy. It will make the car move when it is released.

power

is the amount of *electrical energy* used in a second. It is measured in *watts* (W). See **electrical energy used by an appliance**.

40 watt globe uses less power, heats little and glows dimly.

100 watt globe uses more power, heats up a lot and glows brightly.

predation

occurs when one *animal* (the *predator*) hunts and kills another (the *prey*) for *food*. Predation will affect the *population numbers* of both the predators and the prey, as shown in the graph.

1. Prey numbers increase, providing more food for the predators.
2. Predators have plenty of food so their numbers increase.
3. Prey are eaten so their numbers drop.
4. Predators now have less food, so they die and their numbers decrease.

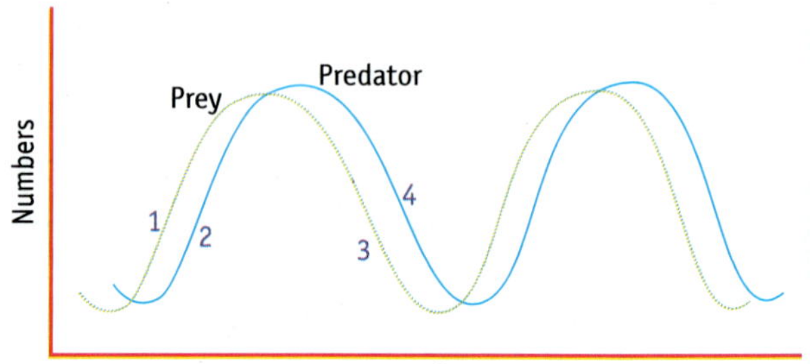

P

predator

is an *animal* that hunts and kills another. See **predation**.

pressure

is the *force* acting on a certain area of a surface.

$$\text{Pressure} = \frac{\text{force}}{\text{area}}$$

Spreading a force over a large area decreases the pressure. See **air pressure; hydraulic machines; pressure and the particle theory**.

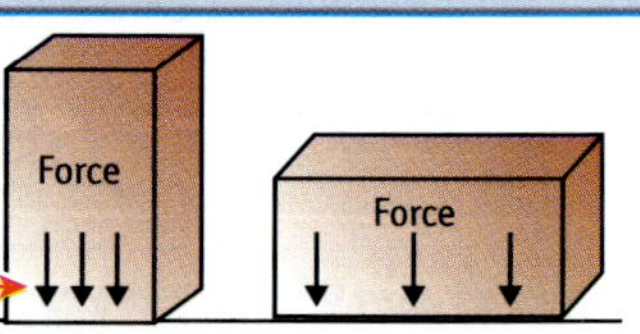

A brick standing on its end applies more pressure to the surface.

Brick pushing down on a small area

Brick pushing on a large area

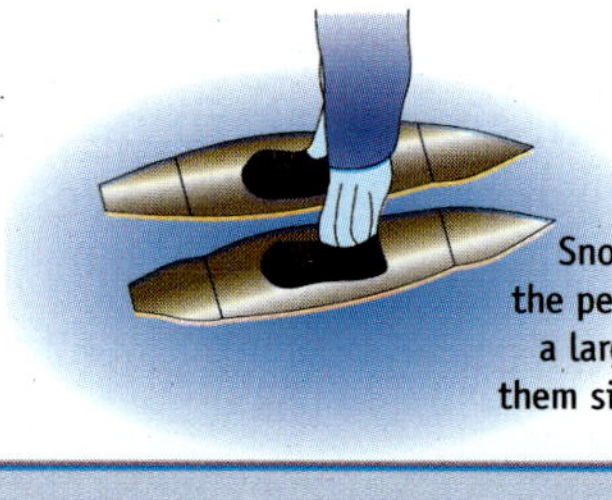

Snow shoes spread the person's weight over a large area and stop them sinking in the snow.

pressure and the particle theory

suggests *pressure* of a *gas* is a measure of collisions of particles with the walls of a container.
See **compression**.

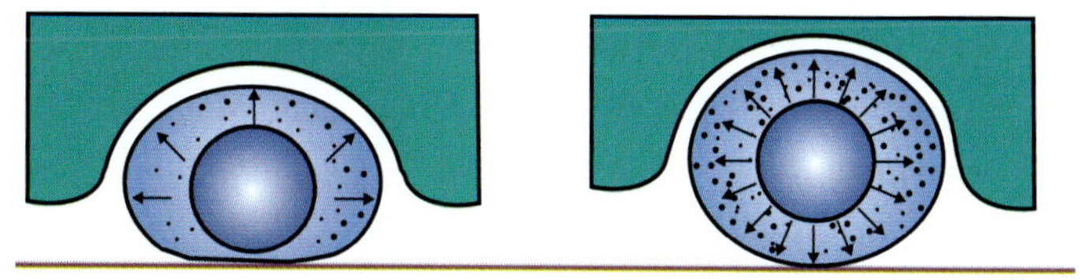

prey

is an **animal** which is killed and eaten by another animal. See **predation**.

processed materials

are natural *materials* that are processed or modified to make them more useful. For instance, wood is converted to timber for building, and *ores* are processed or refined to obtain *metals*. See **processing.** Compare **synthetic materials**.

Examples of processed materials

Cardboard and paper from plants — Timber from trees — Glass from sand — Bricks and cement from the Earth — Metal from ores

processing

is how humans change the *properties* of *materials* to produce items for a particular purpose. Australian Aboriginals made water bags from possum skins, which were strengthened and preserved with tannin from wattle bark. See **processed materials; food preservation; rustproofing; refining.**

producers

are *organisms* that can make their own *food* by a process such as *photosynthesis*. *Plant*s are producers. Producers are at the start of *food chain*s and at the base of *food pyramid*s.

properties

are the characteristics of a substance that make it suitable for a particular use. Properties include appearance, strength, flexibility, *elasticity*, conductivity, *melting* and *boiling points*, *density*, *buoyancy* and solubility. See **property changes; mineral uses; rock use; metal properties**.

Sugar (clear sweet crystals) sinks and dissolves in water.

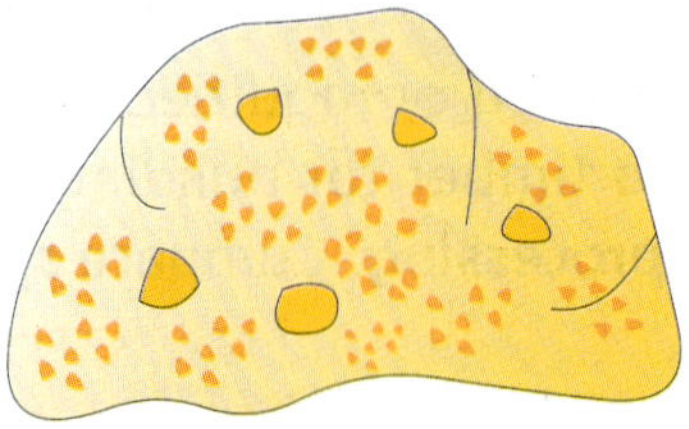

Sponge is spongy, porous and flexible.

Rubber is flexible and does not conduct electricity or heat.

Mat is fibrous.

property changes

result from changing the composition of a substance. This can alter the uses of the substance. For example, adding more sand or gravel to cement makes a weaker concrete, and adding baking powder to flour produces lighter dough than plain flour.

protein

is a *nutrient* needed for formation, growth and repair of body *tissues* (e.g. muscle and *enzyme* formation). It is found in meat, dairy products, cereal grain, nuts and *seeds*.

puberty

is the period when reproductive *organ*s mature and many adult characteristics appear (e.g. breasts and menstruation in female humans, and body hair and deep voice in male humans). See **life span; reproductive system**.

pulleys

are simple *machines* made up of wheels connected with ropes or chains. They are used to change the direction of a *force* or to lift a heavy load with a small effort.

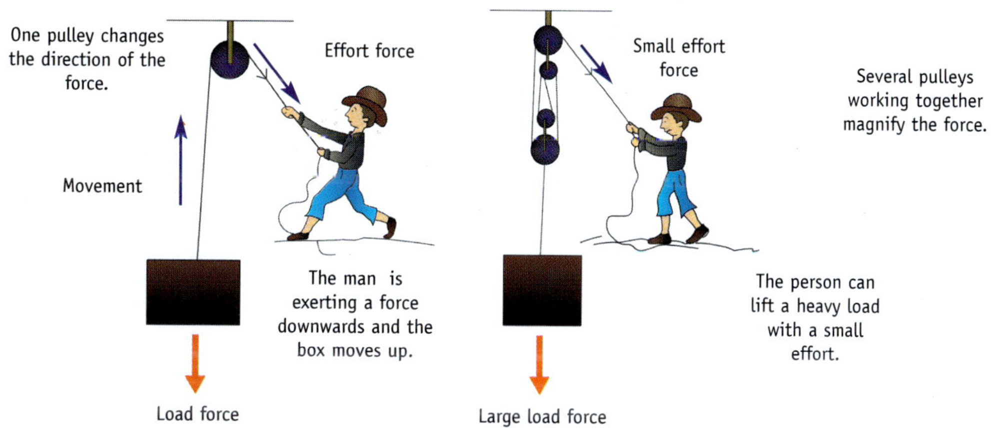

pupa

or chrysalis is a stage in the *life cycle* of some insects. It is usually the stage between the *larva* and the adult and is often found in a protective casing or cocoon. During this stage the larva develops into an adult.
See **metamorphosis**.

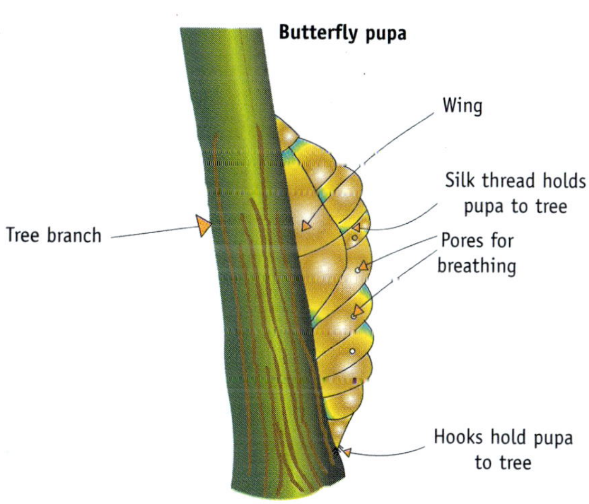

pyramid of numbers

See **food pyramid**.

radiant energy

can travel through empty space. It includes gamma waves, *X-rays*, *ultraviolet* rays, microwaves, radio waves, infra-red rays, *light* and *heat*. The *Sun* is our main source of radiant energy.

radiation of heat

See **heat radiation**.

radio telescope

collects the radio waves emitted by some objects in *space* (e.g. quasers and pulsars) with a large dish. The radio waves are converted to pictures. Radio telescopes can detect objects that are not visible (do not produce or reflect *light*).
See **telescope**.

rain

occurs when fine water particles in the *clouds* combine and become heavy so that they fall. Water frozen slowly forms snow and frozen quickly forms hail.
See **rain gauge**.

rain gauge

measures the amount of rainfall.
A funnel collects the *rain*, which passes into a cylinder where it is measured in millimetres (mm).

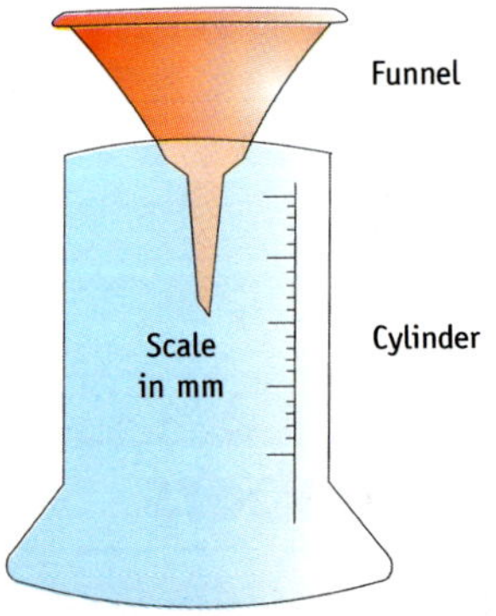

rainbow

occurs because sunlight is made up of different colours. As sunlight passes through a prism or raindrops, the colours are separated and reflected, and are seen as a rainbow.

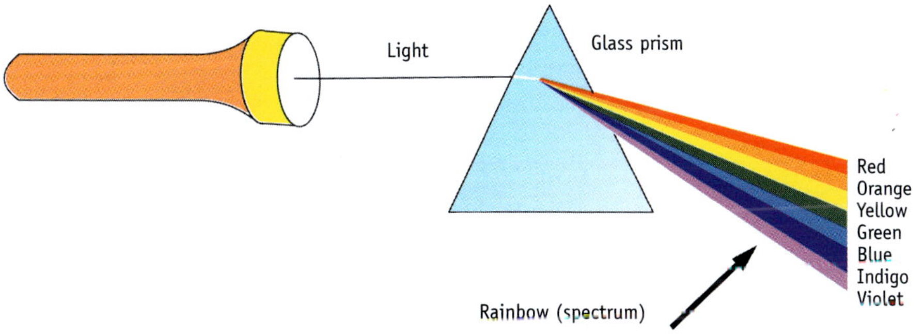

ramp or inclined plane

is a simple *machine*. It is easier to move a heavy object up a ramp than to lift it. A gentle slope requires less effort than a steep one. A staircase is a type of ramp.

Less force is needed to roll the carpet up the ramp than to lift it.

recycle

means to reuse something or remake it so that it can be used again. See **bacteria; waste disposal**.

Red Giant

is a huge red *star*. It forms when the nuclear fuel in a star begins to run out. The star expands, becomes cooler and glows red. Compare **White Dwarf.** See **star, life span.**

refining

is the process of removing impurities from a substance. For example, *ores* are refined to extract *metal*, sugar cane is refined to obtain sugar and *petroleum* is refined to obtain petrol. See **processed materials; fossil fuel uses.**

Iron ore is refined in a blast furnace.

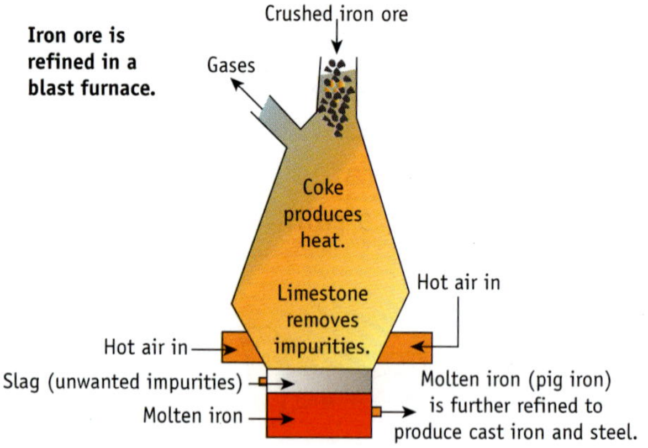

reflection

is bouncing off. *Light*, *heat* and *sound* are reflected from certain hard surfaces. See **mirrors, reflection**.

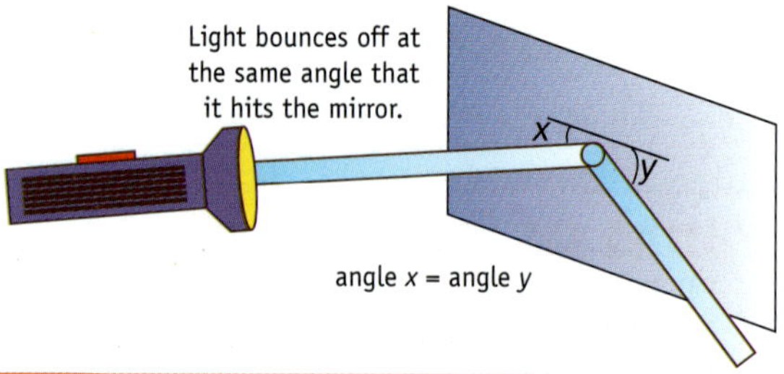

refraction

is the bending of *light* rays as they pass from one substance to another (e.g. from *air* to glass). See **lens**.

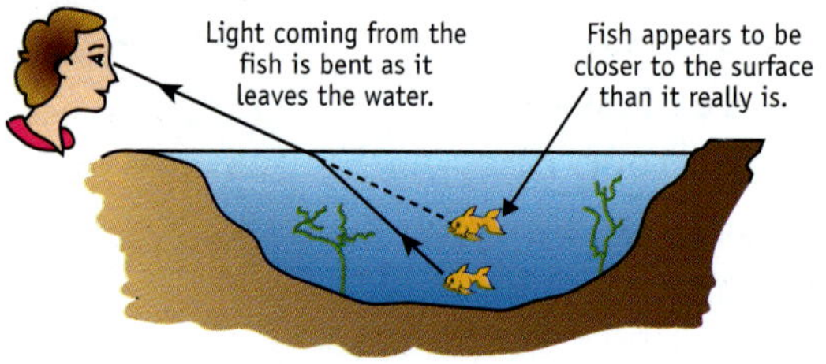

renewable resources

are *materials* that can be readily replaced (e.g. *food*, natural fibres, wood and water). Compare **non-renewable resources**. See **energy sources, renewable**.

Riding a bike uses energy supplied by food which is renewable.

A car uses energy supplied by petrol which is non-renewable.

reproduction

is the process that forms new *organisms*. The young produced are the same kind of organism as the parent (e.g. rabbits produce rabbits). See **asexual reproduction; sexual reproduction**.

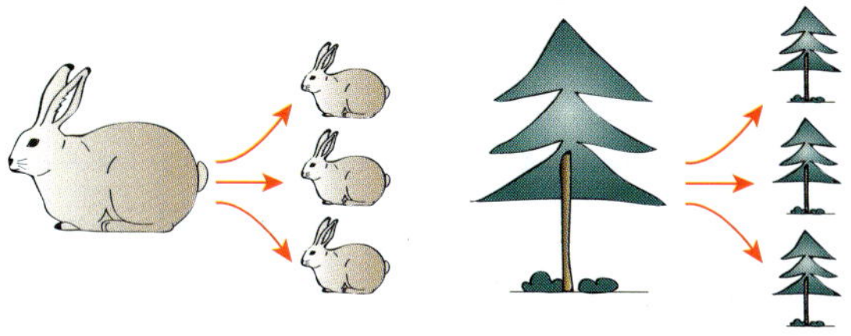

reproduction in flowering plants

occurs in the *flowers*. The male reproductive *cells* in pollen grains are transferred to the female part of the flower by *pollination*. The male cells unite with the *eggs* and an embryo develops inside a *seed*.

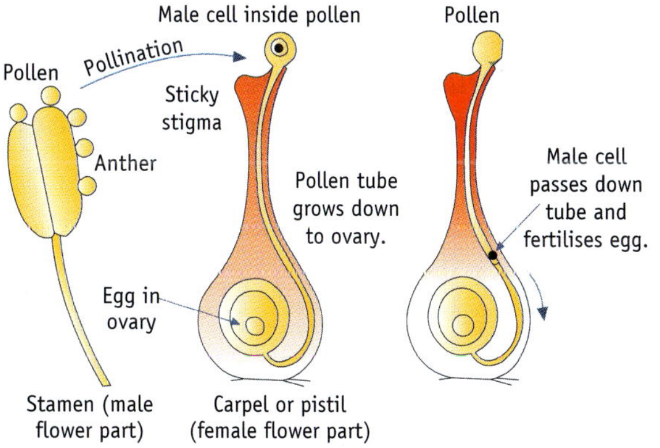

reproductive adaptations

have been made by many *animals* and *plants* to ensure that they can reproduce and the young can survive.

- Fish and frogs: *Eggs* and *sperm* are released into the water and many are eaten. These animals produce large numbers of eggs/ sperm to ensure their survival.
- Birds and lizards: *Eggs* are fertilised inside the animal and a shell is formed to protect the egg. This increases the survival rate.

Frog mating

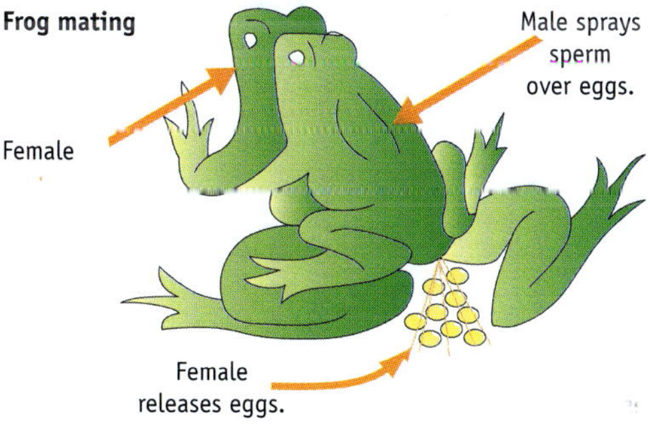

Bird egg

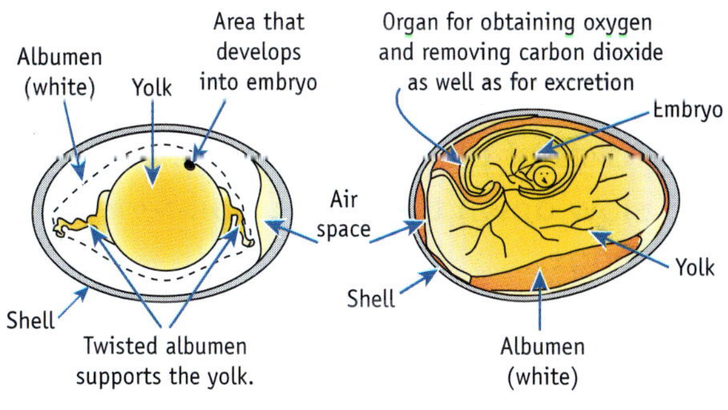

reproductive adaptations (cont'd)

- Mammals: Fertilisation is internal and this increases the chances of successful mating. The embryo develops in the womb or a pouch, where it is protected. See **reproductive system**.
- *Flowering plants*: Most pollen and *seeds* perish. Plants produce huge amounts of pollen and large numbers of *flowers*, *fruits* and seeds to increase the chances of *reproduction* and survival of the *species*.

See **courtship; parental care; reproduction in flowering plants.**

reproductive system

in higher *animals* is made up of *organs* that produce sex *cells* (*eggs* and *sperm*) and, in the female, a place for development of the fertilised egg and the young before birth.

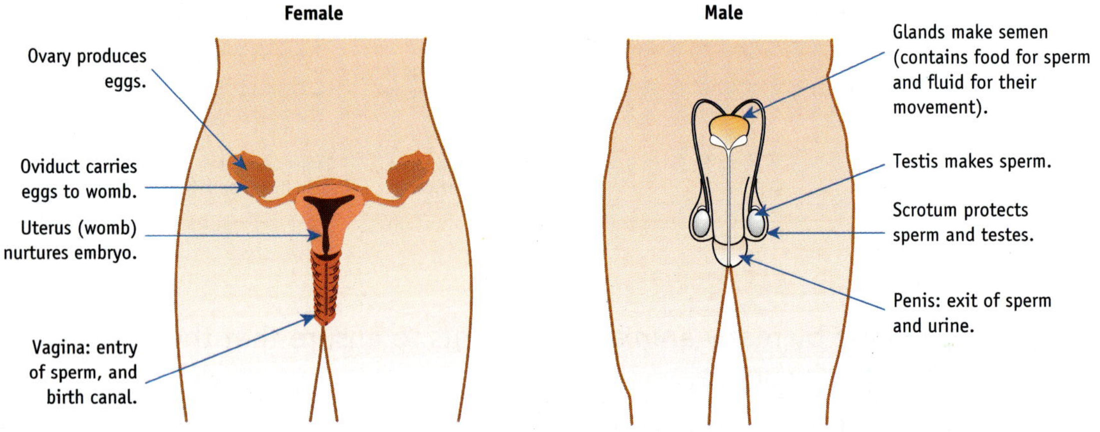

reproductive technology

such as *cloning* and *genetic engineering* are used to select the characteristics of new *organisms*.

resource management

consists of plans set up to:

- slow down the use of *non-renewable resources* and to make supplies last longer;
- reduce harmful waste production;
- repair damage done to the land and the *environment*.

See **sustainable; conservation of non-renewable resources; waste disposal.**

resources

are *materials* that humans use (e.g. *food, building materials*, water, *fuels*). Natural resources come from *living things*, the earth, *air* and water (e.g. wood, *metals*). Many resources are made by humans from natural or raw materials (e.g. *plastics, detergents*). See **processed materials; synthetic materials.**

Resources from the air

Carbon dioxide is used in fire extinguishers and as 'dry ice'.

'Dry ice' keeps ice cream and drinks cold.

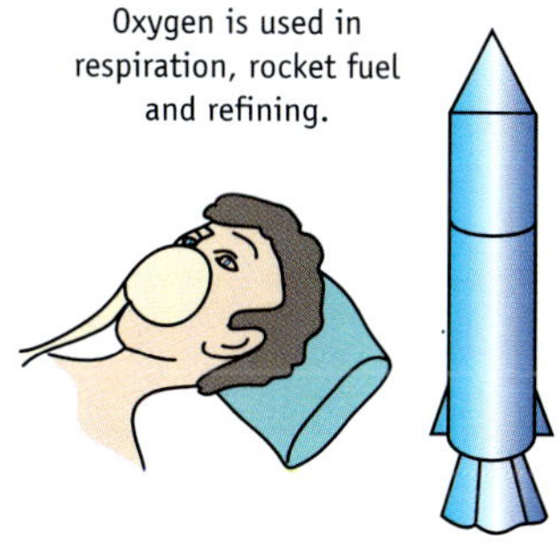

Oxygen is used in respiration, rocket fuel and refining.

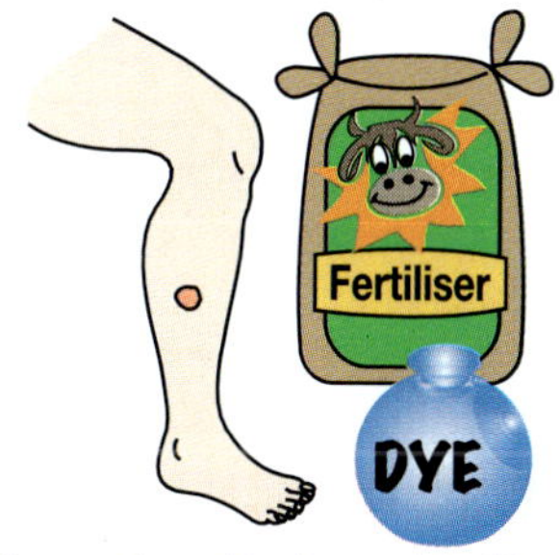

Nitrogen is used in dyes and fertilisers and to remove skin growths.

Resources from the earth

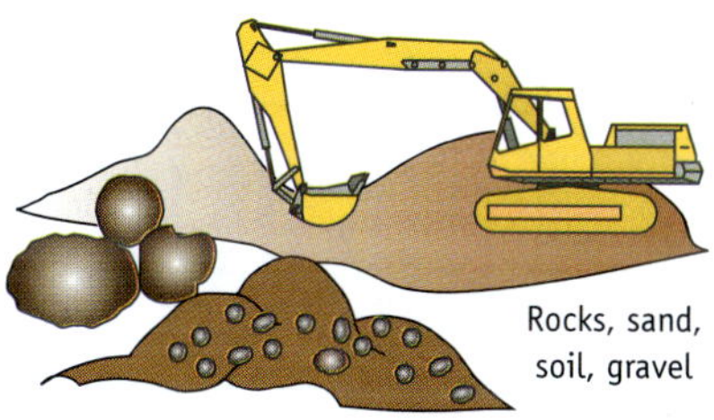

Rocks, sand, soil, gravel

Gold, gems

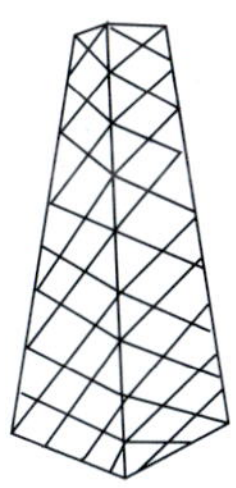

Oil and coal gas

Resources from living things

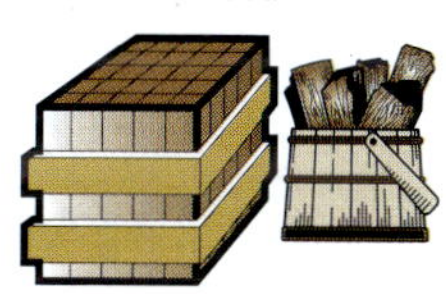

Timber for building and firewood

Fibre

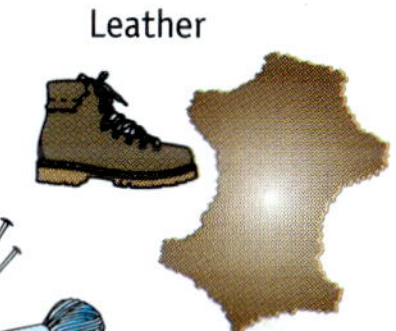

Leather

Food

Flowers

Coal

Oils

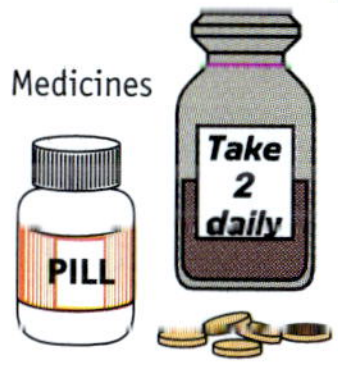

Medicines

Fossil fuels

Resources from water and the oceans

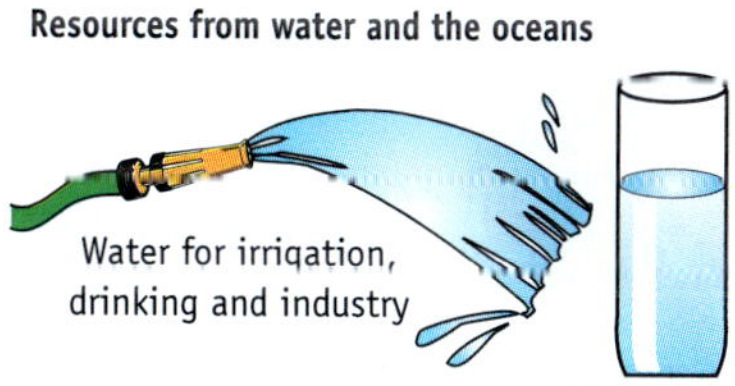

Water for irrigation, drinking and industry

Food

Pearls and shells

Salt

Seaweed for food and fertiliser

respiration

is a series of *chemical reactions* in *cells* of *organisms*. During respiration, cells use glucose (*food*) and *oxygen* to produce *energy* needed for the organism's activities (*life processes*). At the same time, *carbon dioxide* and water are formed.

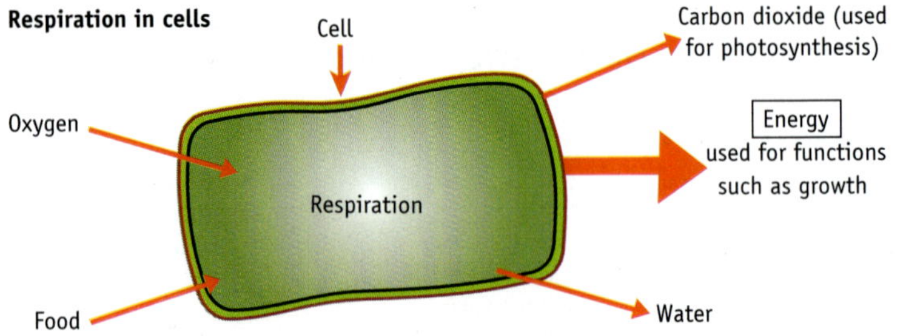

respiratory system

in humans consists of windpipe and lungs. The lungs pass *oxygen* to the blood and remove waste *carbon dioxide* from it. See **breathing; respiration; excretory system**.

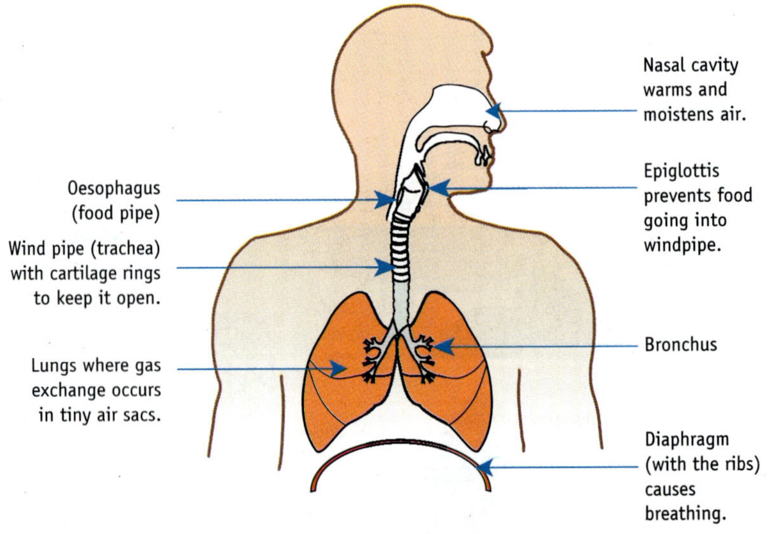

ringbark

is to cut a ring around a tree trunk, in the outer layers of *tissue* (bark). This destroys the *phloem* and stops *food* flow to the *roots* so that the tree dies.

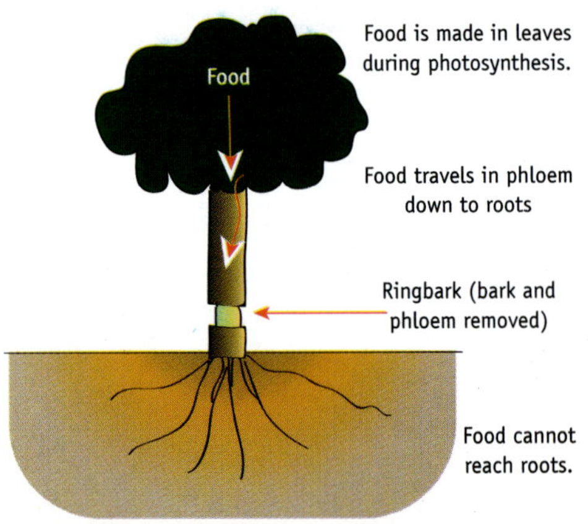

river landforms

consist of hills with V-shaped valleys and waterfalls. As the land flattens out, the river meanders and there are billabongs and terraces. Near the coast are flood plains, estuaries and deltas. See **erosion from water**.

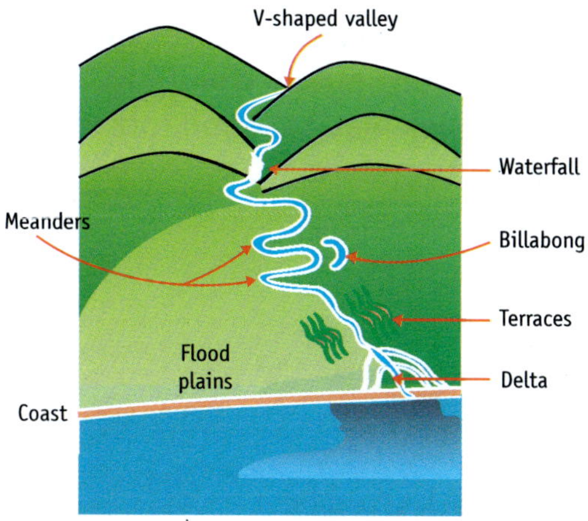

rock

is a *mixture* of *minerals*. Rocks are classified by the way they are formed and their composition. There are three groups of rocks: *sedimentary*, *igneous* and *metamorphic*. See **rock use; rock cycle**.

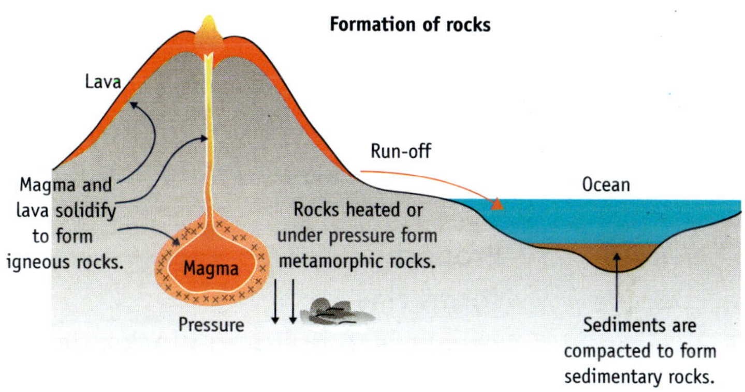

rock age

is determined by the *fossils* or radioactive *elements* in the *rock*. Radioactive elements (e.g. *uranium* or carbon) present in rocks and their fossils break down or decay. The breakdown rate is used to calculate the age of the rock.

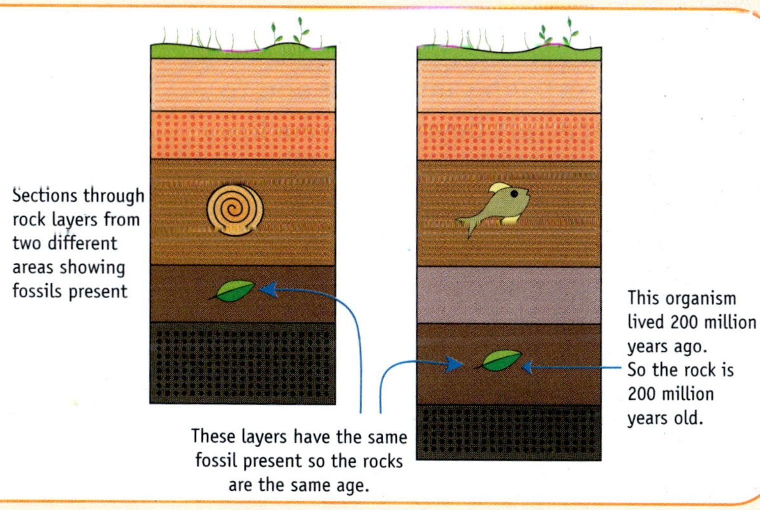

rock cycle

shows that the *rocks* in the *Earth*'s *crust* are continually changing. These cycles may take millions of *years* to complete.

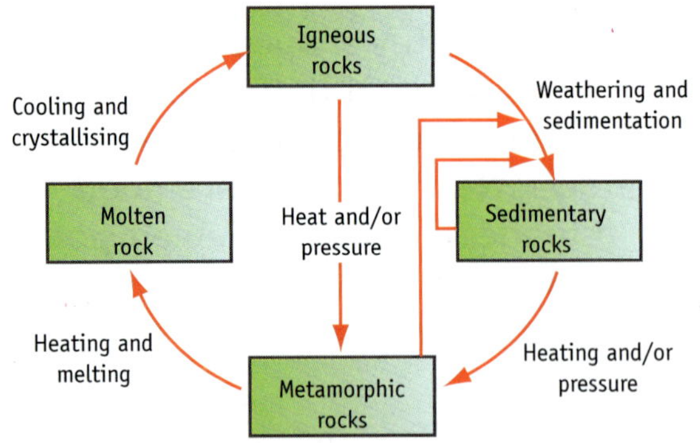

rock minerals

make up the rocks. Common rock *minerals* are quartz, feldspar, calcite, haematite and mica. Compare **ore minerals**.

rock use

is determined by the *properties* of the *rocks*. Rocks used for *building* need to be strong, able to withstand *weathering* and easy to shape (e.g. *sandstone*, granite).

Rock	Property	Use
Shale	forms clay	pipes, bricks
Granite	breaks into building blocks; decorative	buildings; bench tops
Slate	hard and splits into sheets	roofs and floors

rocket

is a strong jet propulsion engine used to place *satellites* into *orbit* and to take *spacecraft* beyond the *Earth*'s *gravity* and into *space*. The first humans into space travelled in rockets made up of stages and modules (e.g. Apollo spacecraft, which landed on the *Moon*).

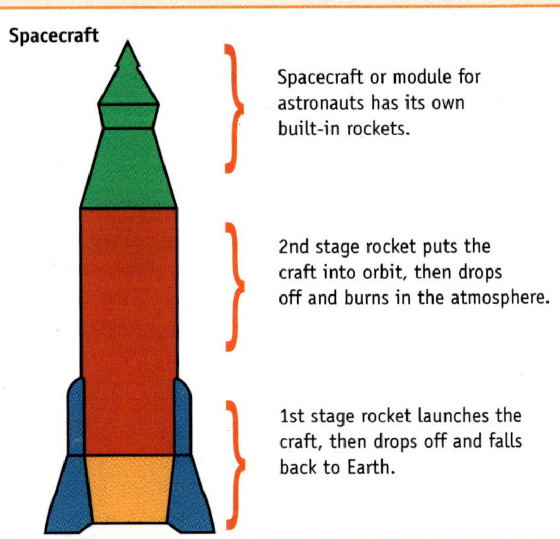

roots

absorb water and *minerals* and anchor the *plant* in the *soil*. They hold the soil together and reduce *soil erosion*. See **physical weathering**.

rotate

means spin on an axis.
The *Earth* rotates once every twenty-four hours. The *Sun*, *Moon* and *stars* appear to move across the sky every twenty-four hours because of this rotation.
See **day and night.**

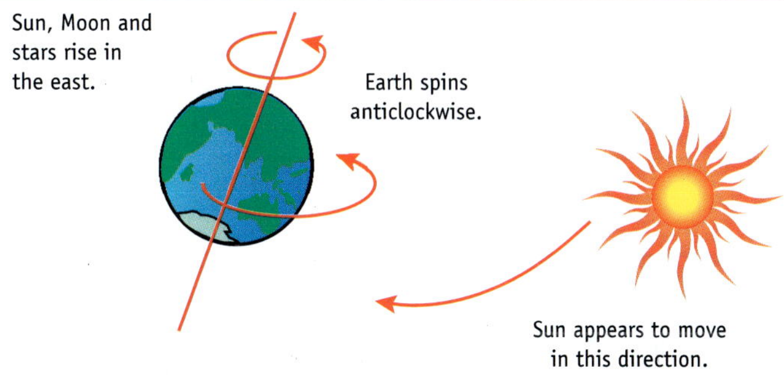

rust

See **corrosion**; **rust proofing**.

rust proofing

involves painting iron to stop water and *oxygen* coming in contact with it. This prevents the *chemical reaction* that forms rust. Coating iron with a more reactive *metal* such as zinc also prevents rusting. This is called galvanising.

salt lakes

form in the desert. Heavy *rains* make lakes that contain *dissolve*d salt from the *weathered rock*s. When the water dries up, a crust of salt remains, forming a lake of salt (e.g. Lake Eyre).

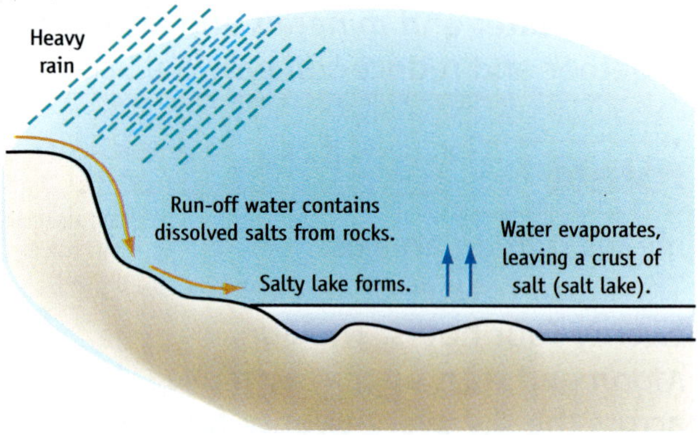

sandstone

is a *sedimentary rock* made of sand particles cemented together. It is formed in oceans, riverbeds and deserts.

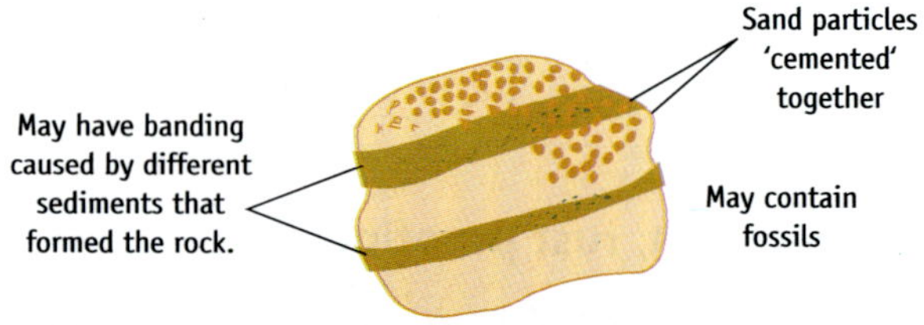

satellite

is a body that travels around another body in *space*. The *Moon* is a satellite that travels around the *Earth*. See **satellites, artificial**.

satellites, artificial

are *spacecraft* that have been made by humans and *orbit* the *Earth* (e.g. Landsat). They make observations, take photographs and collect useful information (e.g. for *weather forecasting*). They also transmit television and telephone messages to other parts of *Earth*. See **satellite**.

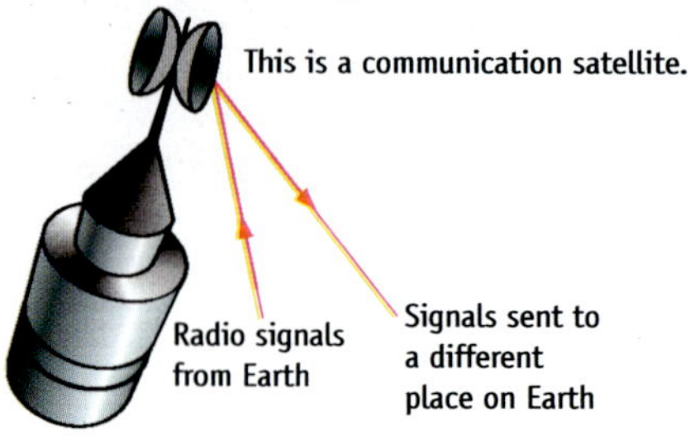

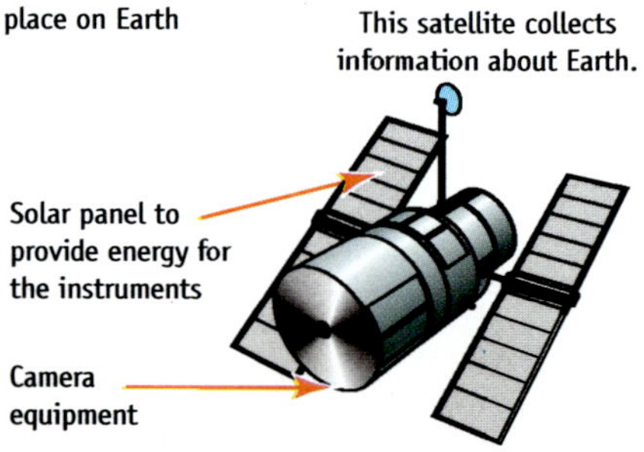

saturated solution

is a *solution* when no more *solid* will *dissolve* in the *liquid*. Increasing the *temperature* of the water will allow more solid to dissolve. Compare **dilute solution; concentrated solution.**

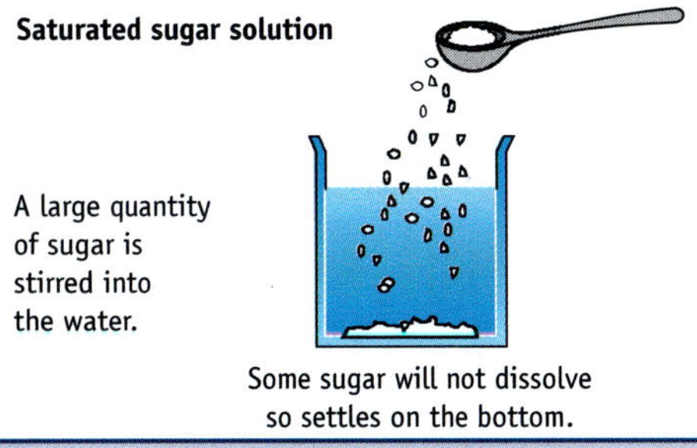

Saturn

is a large *planet* made mainly of *gas*. It is surrounded by a system of rings of ice and rock. The outer surface of gas has belts, ovals and spots (storms). Saturn has more than twenty *moon*s.

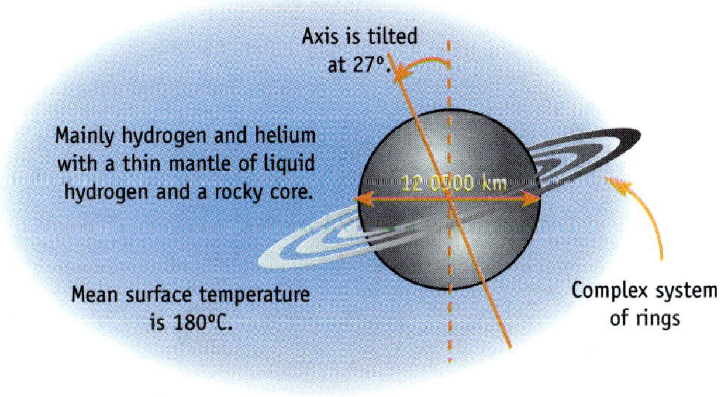

scavenger

is an *animal* that eats dead or decaying animals. Hyenas, crows and ants are scavengers. See **food chain**.

science

The word 'science' comes from the Latin word meaning 'knowledge'. Today it includes the methods used to obtain that knowledge and to solve problems. There are many areas of science and some of these are included in the table:

science	scientist	area of study
astronomy	astronomer	Solar System, celestial bodies and space
biology	biologist	living things
botany	botanist	plants
chemistry	chemist	chemicals that make up the world
geology or Earth science	geologist	rocks, montains, Earth movements and structure
palaeontology	palaeontologist	fossils
physics	physicist	the behaviour of matter (e.g. forces, electricity)
zoology	zoologist	animals

scientific method (fair test)

is a set of steps scientists use to investigate problems and test ideas. See **control; variable**.

Step 1: Observe a problem:

Plants are dying.

Step 2: Suggest a reason (i.e. make an hypothesis):

Perhaps they are not getting enough light.

Step 3: Design and set up an experiment to test this idea:

Experimental group of plants

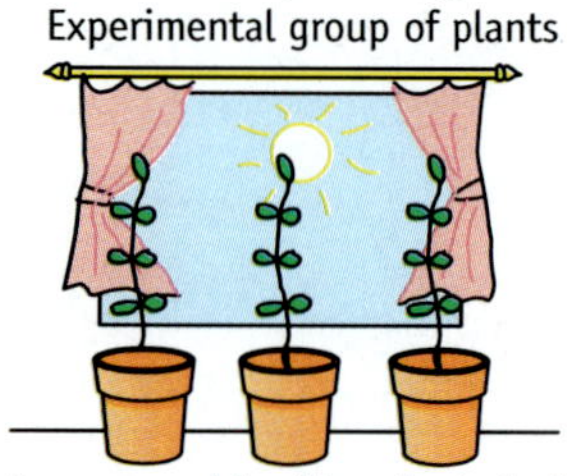

Set up several healthy plants in light.

Control group of plants for comparison

Set up several healthy plants in shade.

Soil, water, temperature are kept the same for both groups of plants.

Step 4: Repeat the experiment

Step 5: Make conclusions from results:

Possible result 1

Alive

Dead

Conclusion: The plants died because they did not have enough light.

Possible result 2

Dead

Dead

Conclusion: Dim light did not kill the plants.
Maybe they did not have enough water (new hypothesis).

seasons

are caused by the *Earth* revolving around the *Sun*. The *Earth* is tilted on its axis. The area that is tilted towards the Sun gets the most direct *heat* and is in *summer*. The area tilted away from the Sun is in *winter*. See **summer and winter**.

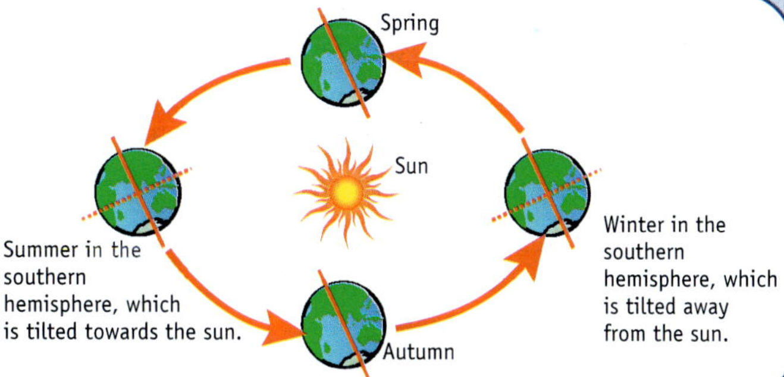

sediment

is a layer of *solid* particles that have settled at the bottom of a *liquid*. Sediments forming in lakes and oceans may eventually form *sedimentary rocks*.

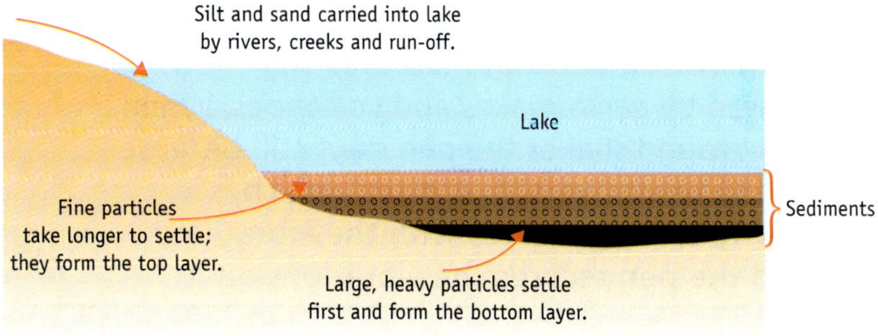

sedimentary rocks

are made from particles of other *rocks*. These particles settle in layers (*sediments*). They are compressed and become cemented together. They may show layers and contain *fossils*. See **sandstone; shale; conglomerate**.

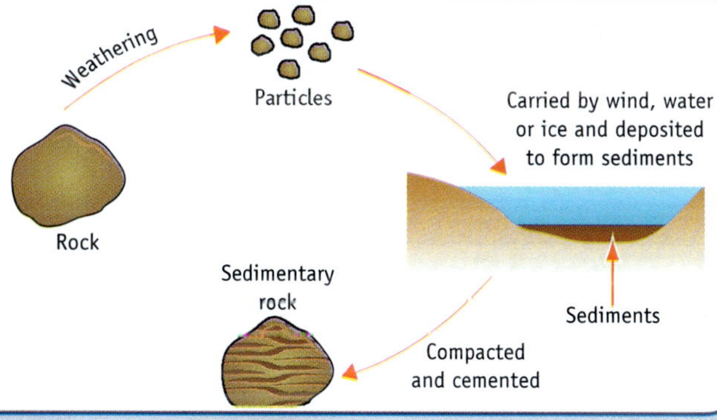

sedimentation and decanting

are methods used to separate mixtures of *liquid* and undissolved (*insoluble*) substances (e.g. water and *soil*). See **water treatment**.

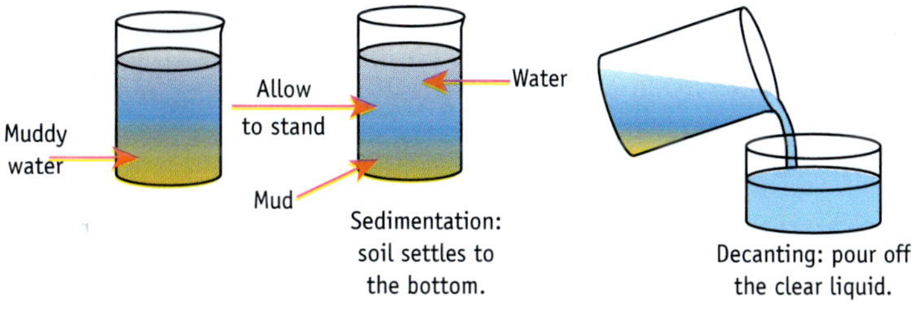

seed

forms after *fertilisation* in flowering and cone plants. It is enclosed in a *fruit*. The seed contains a small *plant* (embryo) that grows when the seed germinates. See **germination**.

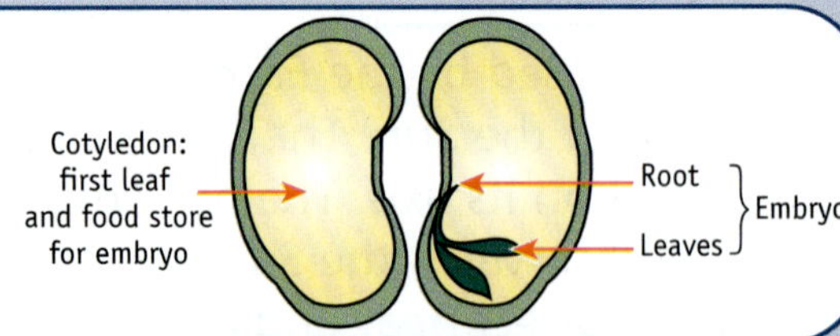

seed capsules

See **fruit**.

seismograph (seismometer)

is an instrument that measures vibrations caused by *earthquakes* and *volcanoes*. When the ground shakes the pen stays still because of the heavy weight. A drum, which is slowly turning, moves with the vibrations and the pen records the vibrations on it.

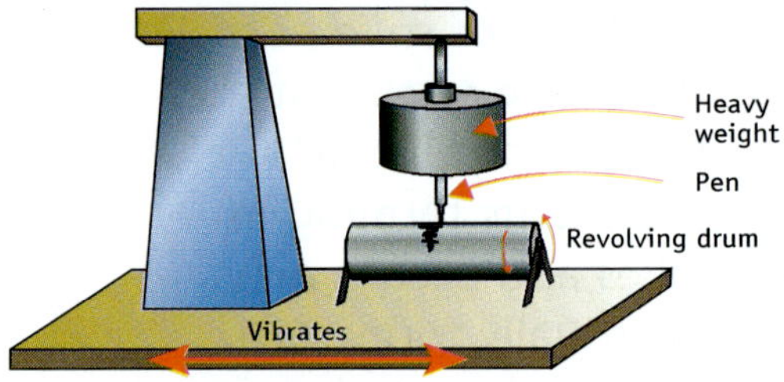

senses

are used by people to observe what is going on around them and to communicate. See **communication**.

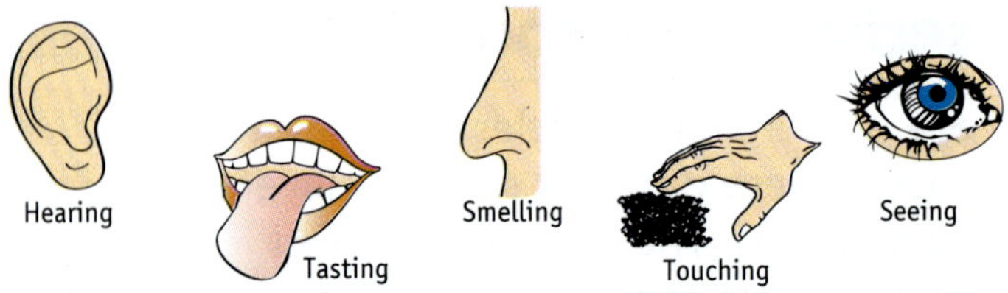

separation methods

are used to obtain useful substances from *mixtures* (e.g. *metals* from *ores*, sugar *crystals* from sugar plant, cream from milk). The substances can be separated because of their different *properties*. Methods used to separate substances include *centrifuging*, *evaporation*, *crystallisation*, *sedimentation and decanting*, *distillation*, *filtration*, *condensation*, *magnetic separation* and *chromatography*.

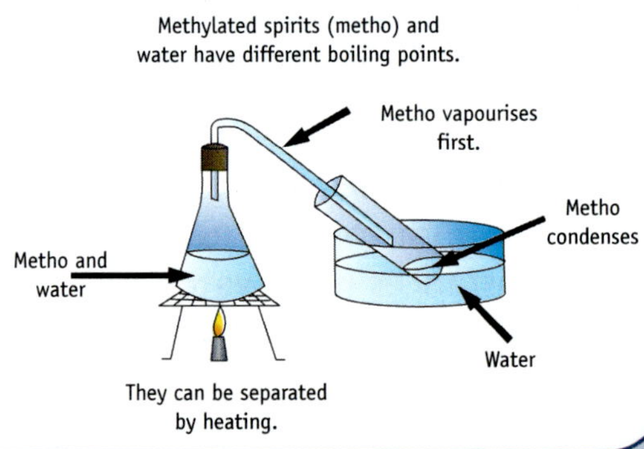

sexual reproduction

occurs when new *organisms* develop after *egg* and *sperm* combine (*fertilisation*). The new individual will have characteristics of both parents. Compare **asexual reproduction**.

shadow

is formed when *light* hits an object it cannot pass through (an *opaque* object). See **shadow stick; sundial**.

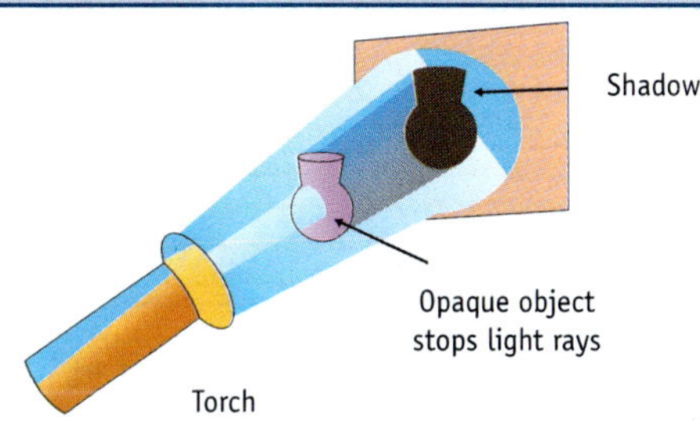

shadow stick

is an upright stick used to produce *shadows* in sunlight. The shadow's length and position changes during the *day*. It can be used to indicate the time. The shadow is long at dawn and dusk and short at midday. See **sundial**.

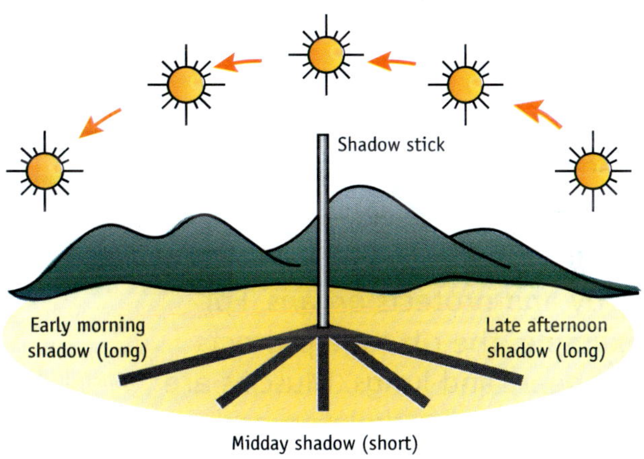

shale

is a *sedimentary rock* made of clay and mud particles. It is formed in still water where these fine particles are able to settle. *Fossils* are often found in shale.

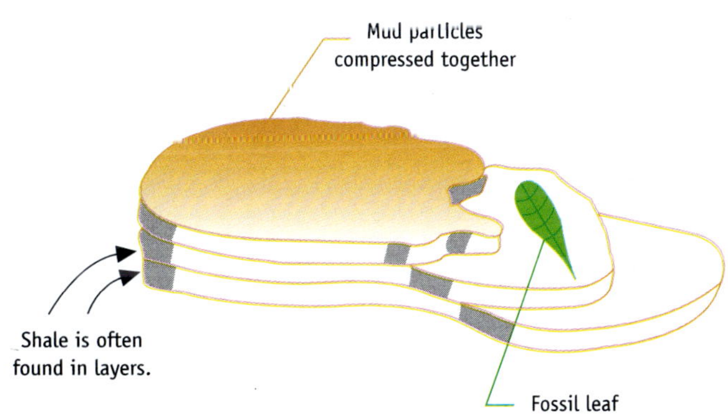

sieving

is one method used to separate substances in a *mixture*. It separates large particles from small ones. It will also separate *solid*s from *liquid*s. See **filtration**.

Sand is separated from rocks by using a mesh or sieve.

Peas are separated from the water by using a strainer.

sink

is to fall gradually. An object that is denser than water (heavier than the same amount of water) will sink. Compare **float**.

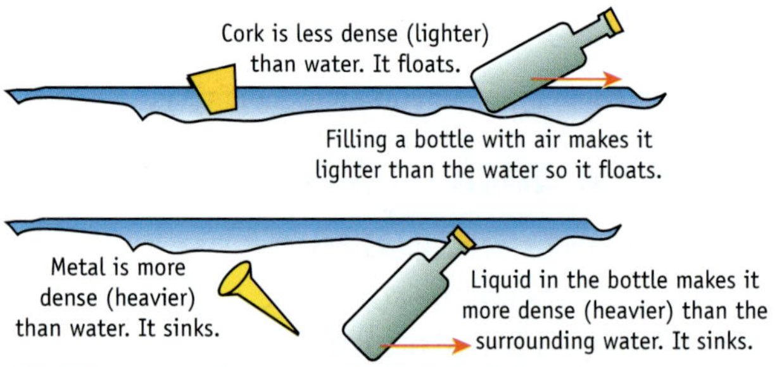

skeletal system

is the skeleton. It supports the body and protects *organ*s. For example, the rib cage protects the heart and lungs. Muscles are attached to the skeleton and together they make the body move. See **levers in the human body**.

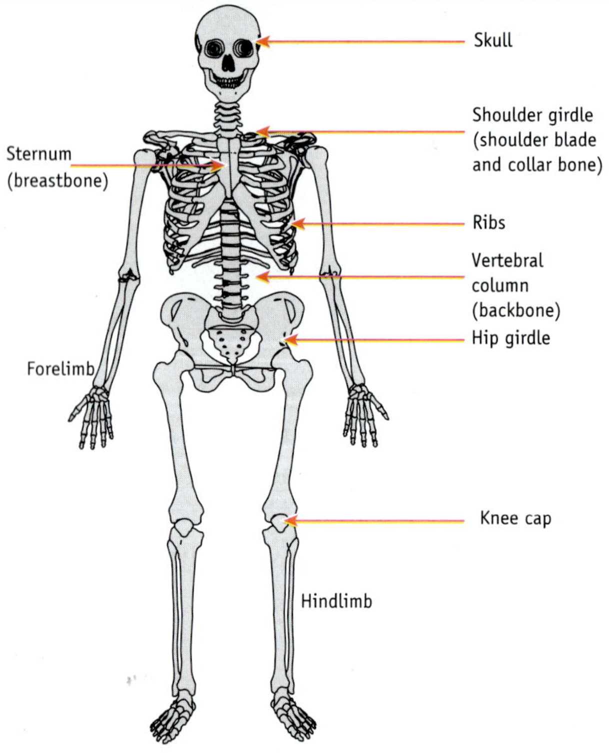

skeleton

See **skeletal system**.

slate

is a *metamorphic rock* formed by *pressure* on *shale*. Layers are sometimes visible in the rock, which splits easily into sheets.

soaps and detergents

are used for cleaning. Soap is made from an alkali and oils or *fat*s. Modern detergents are made from *petroleum* products. Many soaps and detergents have polluted the *environment*. *Biodegradable* soaps and detergents are now being manufactured. *Microorganism*s change them to harmless products.

soil

consists of *weathered rock* particles (sand, silt and clay) and decaying *plant* and *animal matter* (humus). The diagram shows a cutting through soil layers, showing the different stages in soil formation. See **soil composition**.

Soil profile

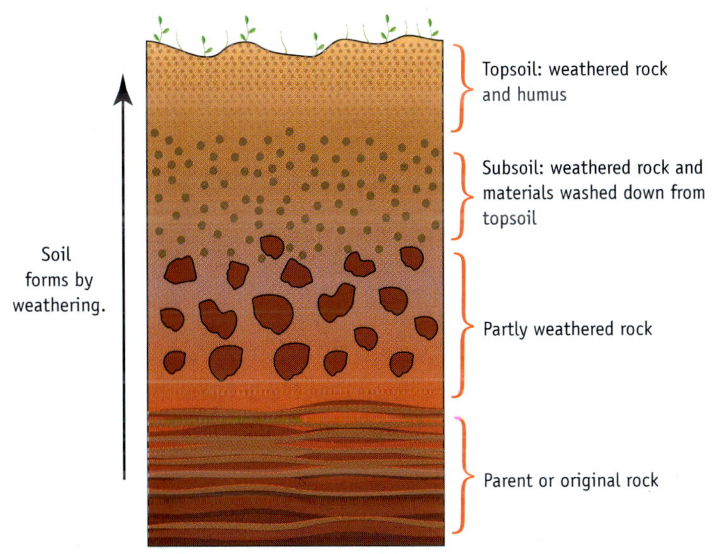

soil composition

is the amount of clay, silt and sand present. *Soil* is classified by its composition. For example, loams are rich soils with about equal amounts of sand, silt and clay with humus. Clay, silt and sand give the soil its texture and make it *porous* (allowing air spaces). See **soils, water retention**.

soil conservation

is protecting or preserving the *soil* against *erosion* and depletion of *nutrients*. Erosion is reduced by growing *plants*, by reducing grazing and by farming on terraces and in contours around a hill rather than ploughing down it. Nutrients can be replaced by *composting* and mulching. See **soil erosion.**

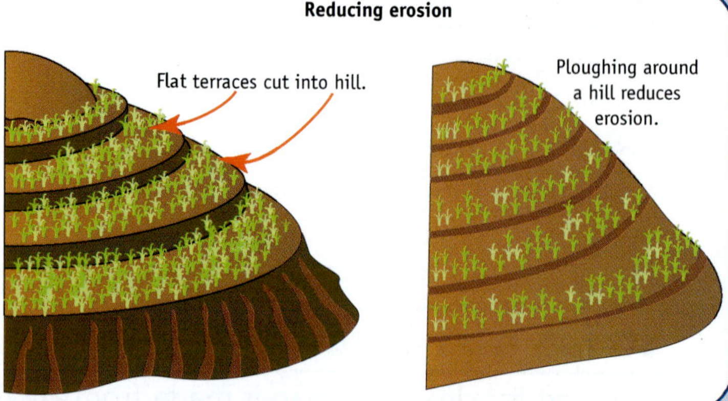

soil erosion

occurs when rich topsoil is washed away, leaving behind land with poor *soil* and trenches. It is caused by removing *plants* from the land and by overgrazing.
See **soil conservation**.

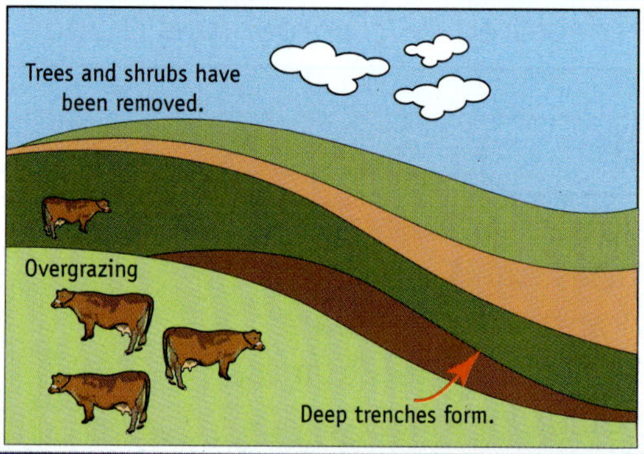

soil texture

depends on the amount of sand, silt or clay present. Texture can be felt by rubbing the *soil* between the fingertips. Soils with a coarse texture have a lot of sand and feel rough. Soils with a fine texture contain mainly clay.

soils, water retention

is better in some soils than others. Sandy soils are very porous and do not hold water. Clay soils hold onto water and tend to become waterlogged. See **soil composition**.

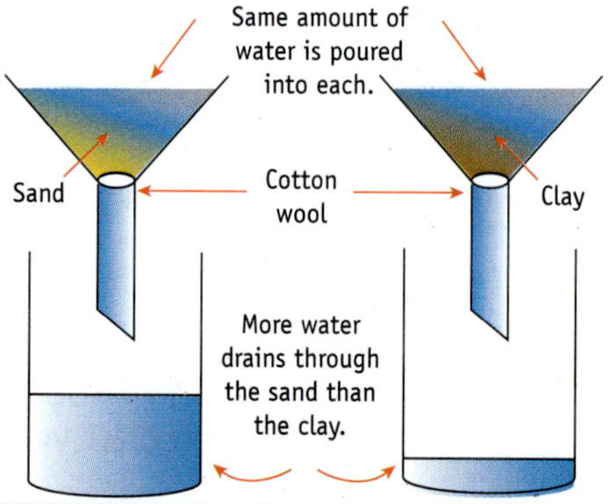

solar energy

is *radiant energy* from the *Sun* (e.g. *heat, light and ultraviolet rays*). It is the main source of energy on *Earth* and is responsible for natural activities such as *photosynthesis*, *wind* and *rain*. Solar energy can also be used to heat the household water supply and can be changed to *electricity* by solar cells. See **energy sources, renewable; energy transfer; energy transformation in living things**.

Solar water heater

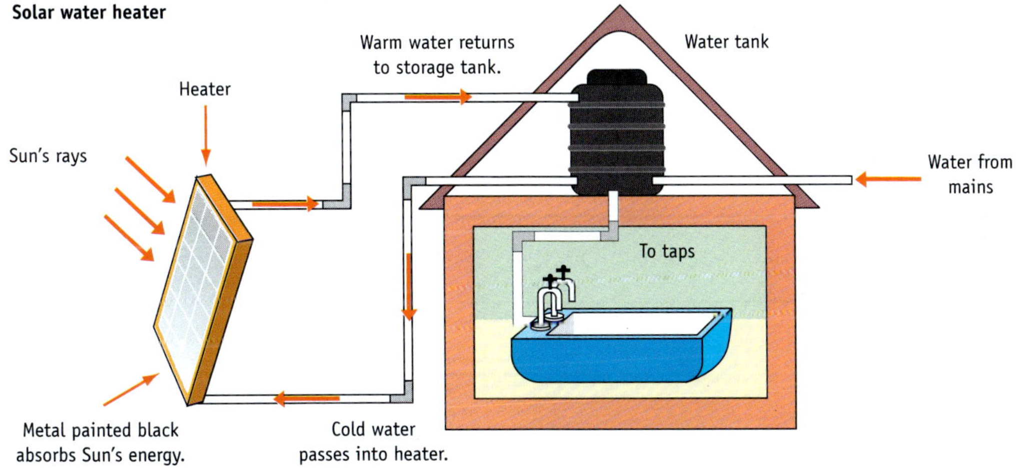

Solar System

is made up of the *Sun*, the nine *planets* and their *moons*, and the *asteroids* and *comets* that all move around the Sun.

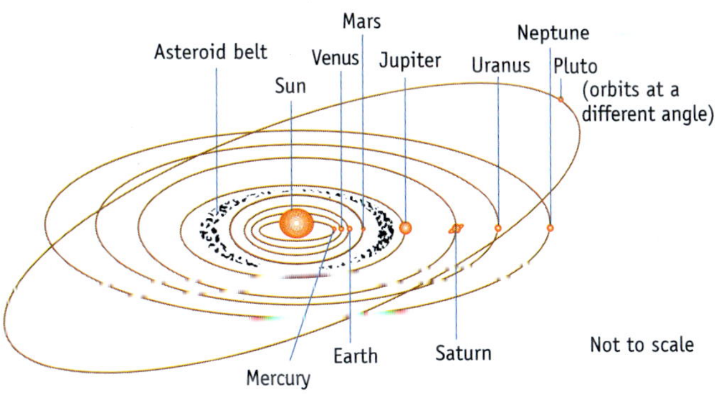

solid

is one of the three *states of matter*. The particles in a solid cannot move freely so solids have a definite shape and size, or *volume*. See **particle theory of matter**.

soluble

describes a substance that *dissolves* in another.
A substance that does not dissolve is *insoluble*.
For example, sugar is soluble in water, but oil *floats* on water and so oil is insoluble in water.
See **solution**.

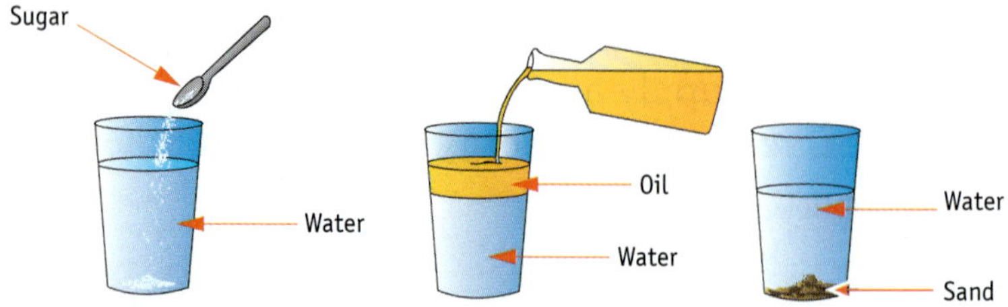

Sugar dissolves in water. It is soluble.

Oil floats on water. It is insoluble.

Sand settles at the bottom of the water. It is insoluble.

solute

See **solution**.

solution

is formed when a substance *dissolves* in a *liquid*. The liquid is called a *solvent*. The dissolving *solid* is called a solute.
See **dilute solution; concentrated solution; saturated solution**.

Sugar solution

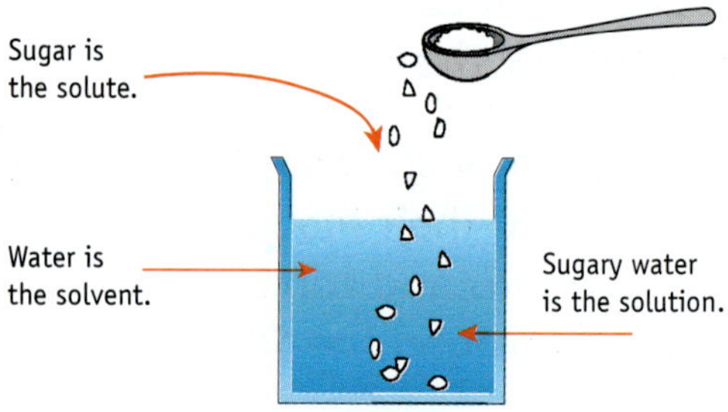

Solutions made with water are called aqueous solutions.

solvent

is a *liquid* that *dissolves* substances. Water dissolves many substances and forms aqueous *solutions*. Turpentine is a solvent used to dissolve oil paint; acetone in nail polish remover is a solvent that dissolves nail polish.

Turpentine

Nail polish remover (acetone)

sound

See **sound energy**.

sound energy

is caused by vibrations. These vibrations make *air* particles move in waves, which travel to your ear. Sound will not travel through a *vacuum*.
See **pitch; loudness; energy transfer; energy transformation**.

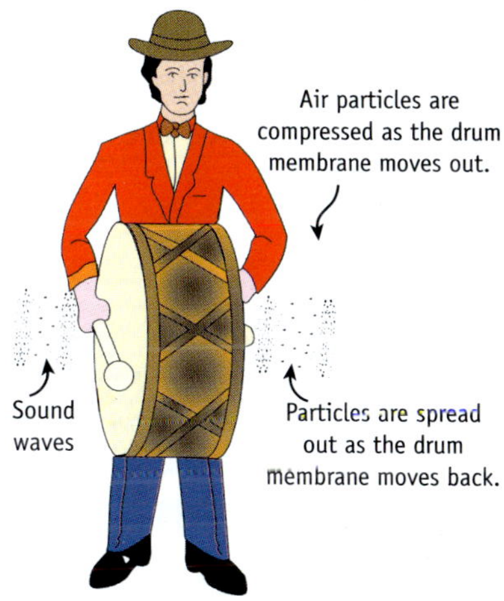

sources of energy

See **energy sources**.

space

is the part of the *Universe* beyond the *Earth*'s *atmosphere*.

space probes

are unmanned *spacecraft*. They travel into *space* and collect information about space without interference caused by the *atmosphere*. The probes carry instruments to measure conditions and send this information back to *Earth*. For example, Voyager 2 travelled to *Saturn* and *Jupiter*.

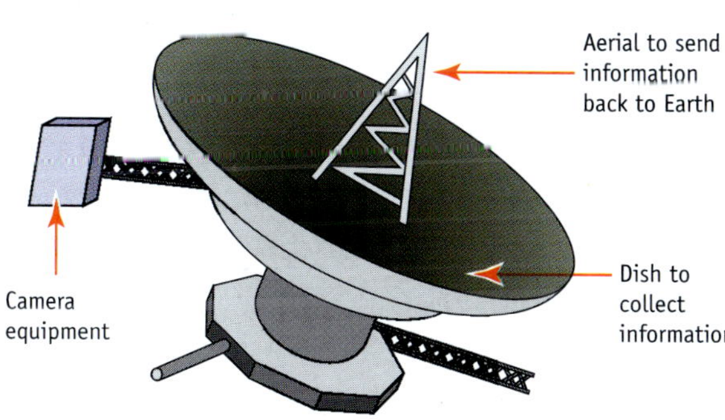

space shuttle

consists of a launch *rocket* and a manned *spacecraft*, which returns to *Earth* like a plane and can be reused. The space shuttle is used to take *satellites* and *space stations* into *space*, carry crews and equipment to and from space stations and take crews out to repair satellites.

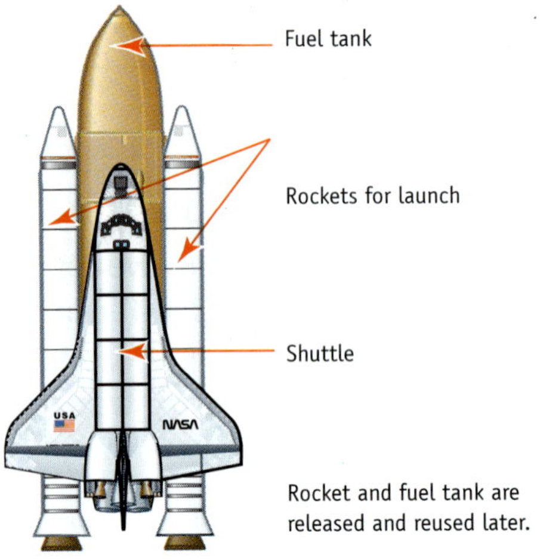

Rocket and fuel tank are released and reused later.

space stations and laboratories

are manned by scientists who make observations and can carry out research and *experiments* in *space*. Examples are Skylab, Mir and the International Space Station (ISS).

space technology

is the *technology* developed for exploring *space* (e.g. *telescopes* and *spacecraft*). The items are used for collecting information and carrying out research.

spacecraft

include *rockets*, *satellites*, *space probes*, *space shuttles* and *space stations and laboratories*. They are used to explore and collect information about *space*.

species

is a group of *living things* that can breed and produce young in nature. For example, lions and tigers do not naturally interbreed; they belong to different species. See **variations within a species**.

Do not naturally interbreed.

speed

is the distance travelled in a certain time. It is measured in metres per second (m/s) or kilometres per hour (km/h).

$$\text{Average speed} = \frac{\text{distance travelled}}{\text{time taken}}$$

The girl and her dog take 10 seconds to travel 10 metres. They have a speed of 1 metre per second.

sperm

is the male reproductive *cell*. In higher *animals* sperm is formed in the testes. See **fertilisation**.

star

is a glowing ball of *gas*. It forms from a cloud of dust and gas, which contracts and is held together by *gravity*. Nuclear reactions occur in stars, releasing *heat* and *light*. Our *Sun* is a star. See **Red Giant; White Dwarf; star, life span**.

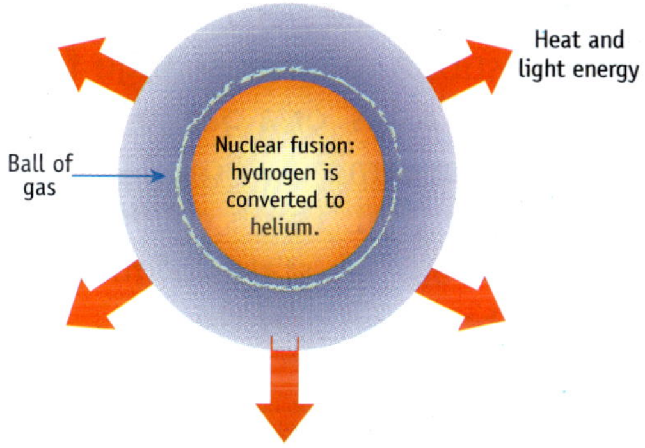

star, life span

stars last for thousands of millions of years and change with age. At the end of its life, the star may become a *White Dwarf* which becomes cold and fades. Some stars erupt with a violent explosion (super nova) to form a *nebula*. Sometimes very large stars contract to form a *black hole*.

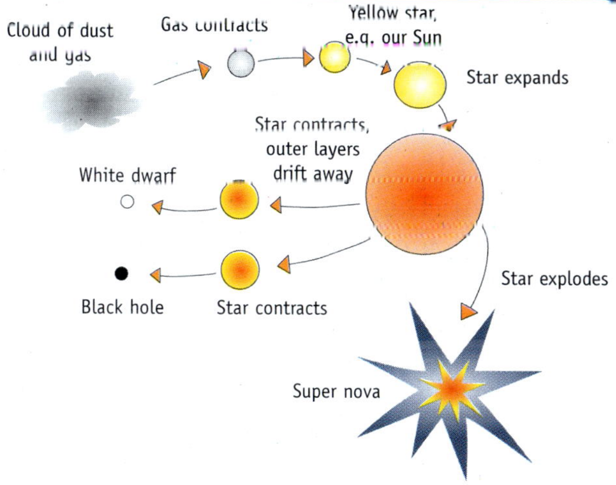

state of matter

are the three states in which matter exists: *solid*, *liquid* and *gas*. For instance, water exists as a liquid, as solid ice or as a gas, when it is steam. See **state of matter, change; particle theory of matter**.

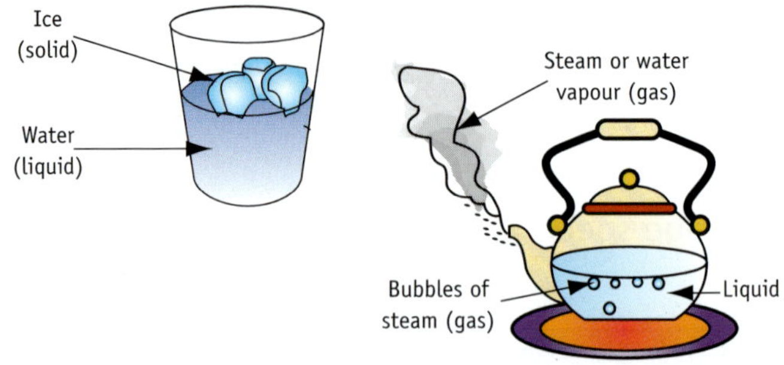

state of matter, change

is the change of state that occurs when a *solid* becomes a *liquid*, a liquid become a *gas*, a gas forms a liquid or a liquid becomes a solid. A few substances will change from a solid to a gas, or a gas to a solid, without becoming a liquid (e.g. dry ice, iodine). This is called sublimation. See **states of matter; forces between particles**.

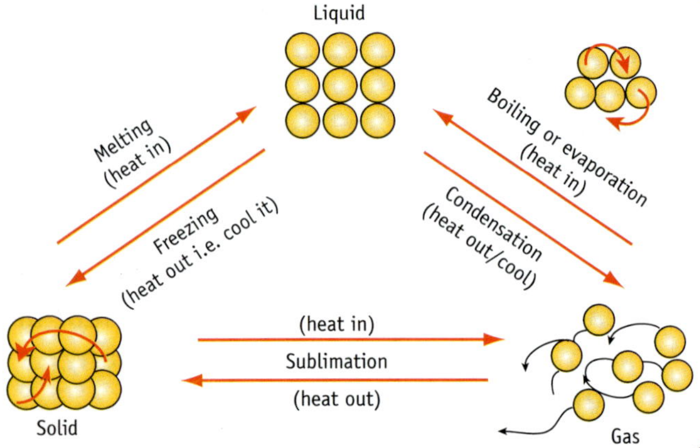

static electricity

is the *charge* collecting on objects. The charge causes attractive and repulsive *forces*. See **electrostatic force; charging an object**.

stems

support the *plant* so that its *leaves* receive sunlight. They carry water and *minerals* in *xylem tissue* from the *roots* to *cells* where the *food* is needed. They also carry *food* that has been made in *phloem* tissue in the leaves to cells.

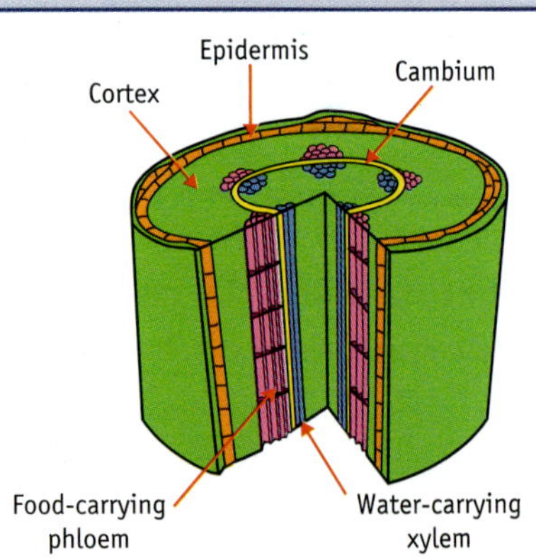

stomates

are pores in the surfaces of *leaves* (and sometimes *stems*). They allow water to evaporate and *carbon dioxide* and *oxygen* to pass into and out of *plant*s. See **transpiration; respiration; photosynthesis**.

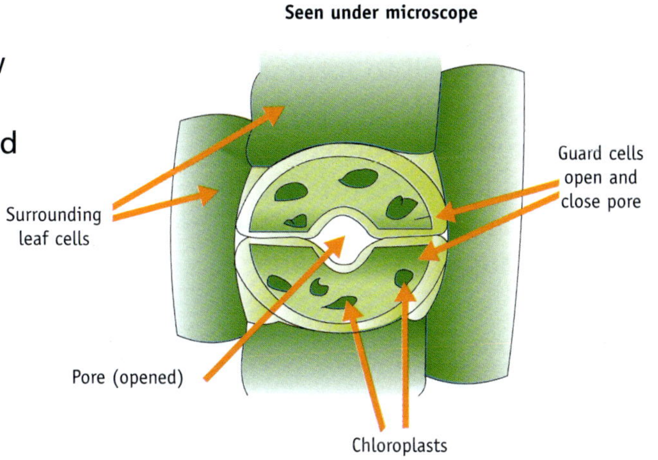

strength and shapes

See **material shaping**.

sublimation

See **state of matter, change**.

summer and winter

are determined by the angle at which the *Sun*'s rays hit the *Earth*. In summer, the southern hemisphere is tilted towards the Sun and is closer to it. The Sun's rays are concentrated in a smaller area and the ground heats readily. In winter, the southern hemisphere is tilted away from the Sun. The rays hitting the Earth are spread over a large area and so heating is less. See **seasons**.

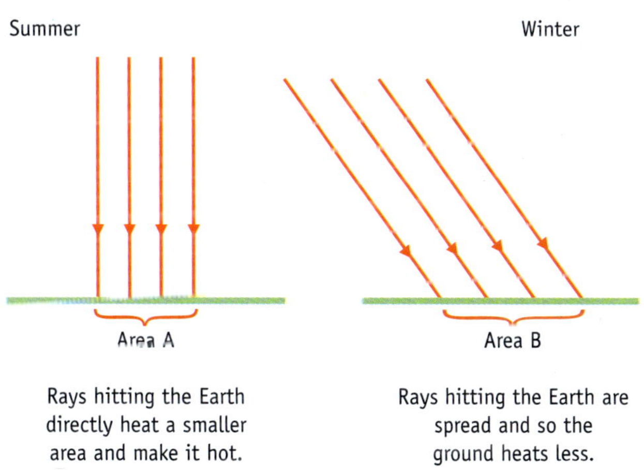

Area A is smaller than Area B.

Sun

is a yellow *star* and will eventually swell to form a *Red Giant*. It is made of hydrogen and helium *gas*. Nuclear reactions in the centre of the Sun convert hydrogen to helium and produce *heat* and *light energy*.

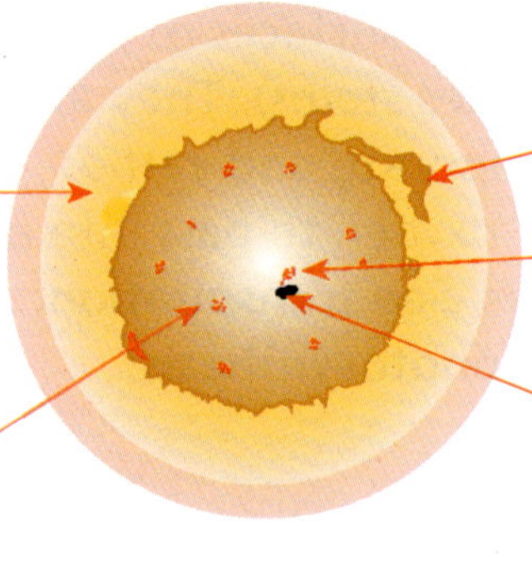

sundial

is an instrument that indicates the time by using a *shadow* produced by the *Sun* and a pointer. The shadow falls on different parts of the dial at different times of the *day*. See **shadow stick**.

survival features

have evolved to help *living things* survive. For example, birds have evolved beaks that enable them to eat different *food*s. See **adaptations**.

suspension

is a *mixture* made up of a *liquid* containing *solid*, *insoluble* particles that will eventually settle on the bottom. Compare **solution**.

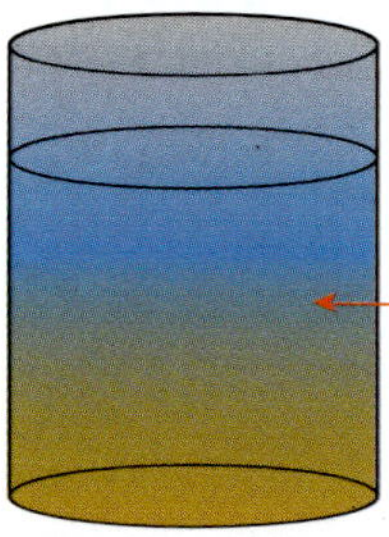

sustainable

is making supplies last longer by methods such as *recycl*ing, reusing and using alternatives. Aboriginal Australians cared for the land so that it could recover between periods of use. Their nomadic lifestyle gave time for *food* supplies (e.g. fish, *plant*s) in an area to regrow. See **resource management**.

switch

operates a *circuit*. Turning an electrical switch off makes a gap in the circuit so that *current* will not flow. This is called an open circuit. Turning a switch on closes the gap and allows *electricity* to flow. This is a closed circuit.

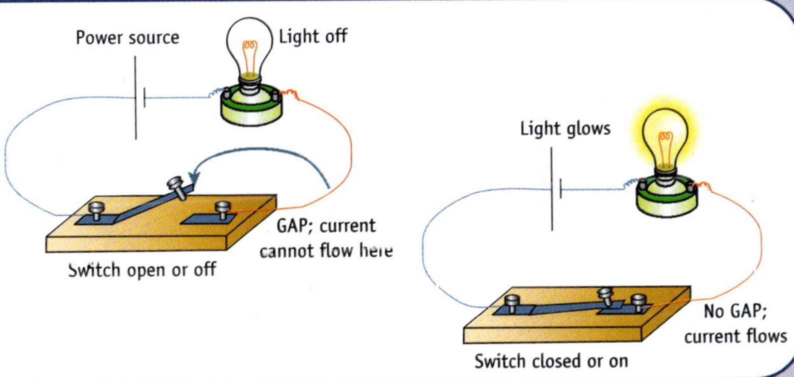

symbiosis

is a close association between *organisms* of different *species*. See **ecosystem interactions**.

Staghorn fern perched on tree trunk helps it to get light.

Sucker fish attach to shark and eat its food scraps.

symbols

See **electrical symbols**; **element symbols**.

synthetic materials

are artificial *materials* made by humans from natural or raw materials (e.g. *plastics* and *detergents* are made from *petroleum* products). They are often used instead of natural materials because of their different characteristics, or *properties*. See **impact**.

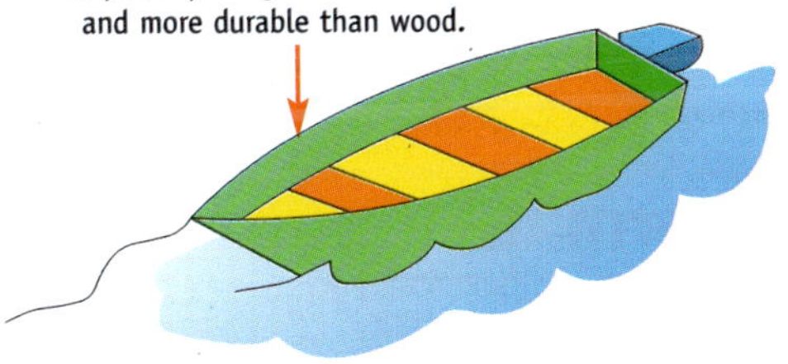
Fibreglass boat (glass embedded in plastic) is lighter than metal and more durable than wood.

technology

refers to the tools and techniques used to carry out a plan. Sticks to dig up yams, string bags made from bark fibre and fishing hooks made from shells are examples of technology. Modern technology includes complex *machines*, such as cars, computers and *satellites*.

Technology used by the traditional Australian Aborigines:

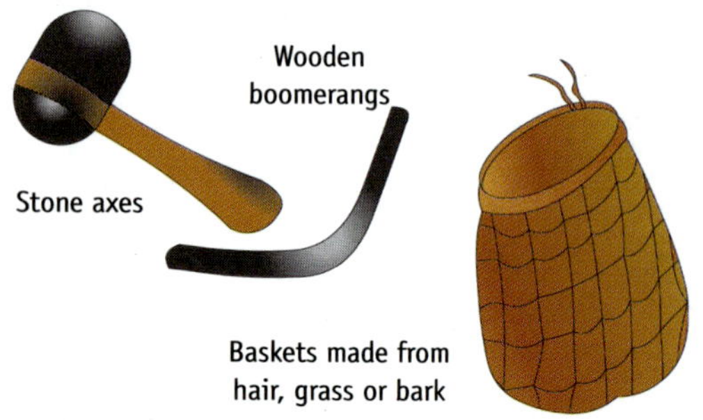

Modern technology:

telescope, optical

is used to observe objects in *space*. It can magnify objects and also collect *light* from very faint objects. Reflecting telescopes collect light with *mirrors* and refracting telescopes collect light with *lenses*. See **radio telescope; Hubble Space Telescope**.

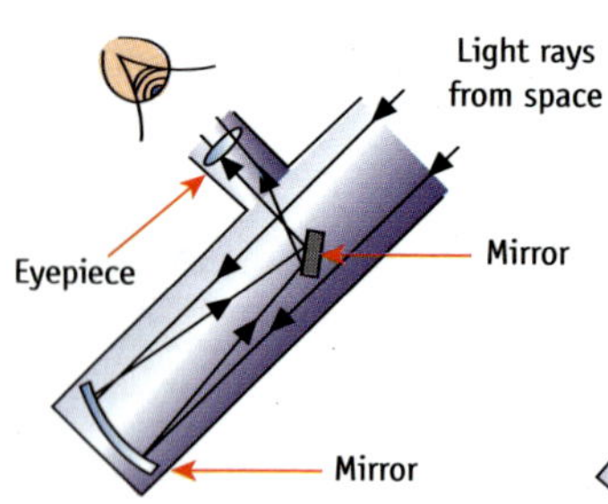

Reflecting telescope

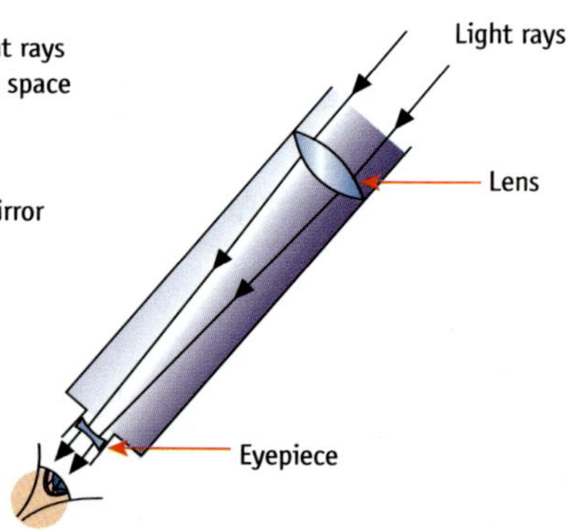

Refracting telescope

temperature

indicates the hotness of a substance. It is measured with a *thermometer* in degrees Celsius (°C). In some countries it is measured in degrees Fahrenheit (°F).

termite

See **colony**.

terrestrial

means living or growing on land. Terrestrial *environments* include deserts, mountains, grasslands, farms and cities.

thermometer

is used to measure the hotness of a substance. Most thermometers are glass tubes containing mercury.

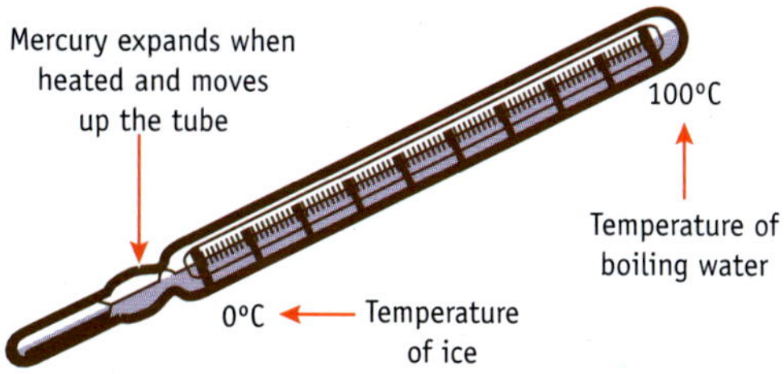

thermos flask

is a good example of *heat conservation*. It is designed to reduce heat loss by *conduction*, *convection* and *radiation*.

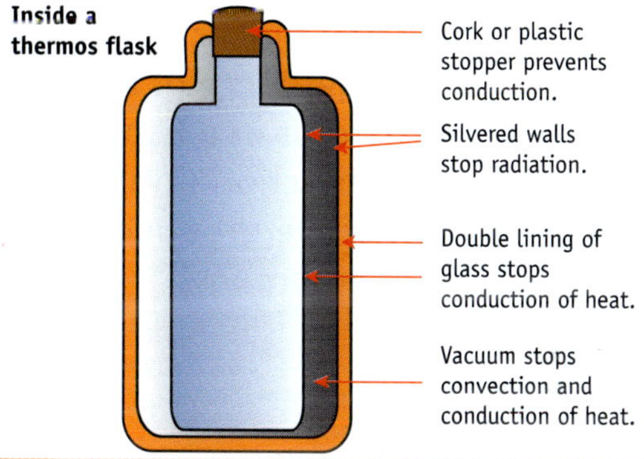

tidal energy

uses the moving water from large *tides* to rotate *turbines*, which drive *generators* and produce *electricity*. See **energy sources, renewable**.

tides

occur as the height of the *ocean* rises and falls twice a *day*, causing high and low tides. A high tide results when the *Moon*'s gravitational pull moves the *Earth*'s water towards it. Low tide occurs at places where the water is moving away. The *Sun* also has a small effect on tides.

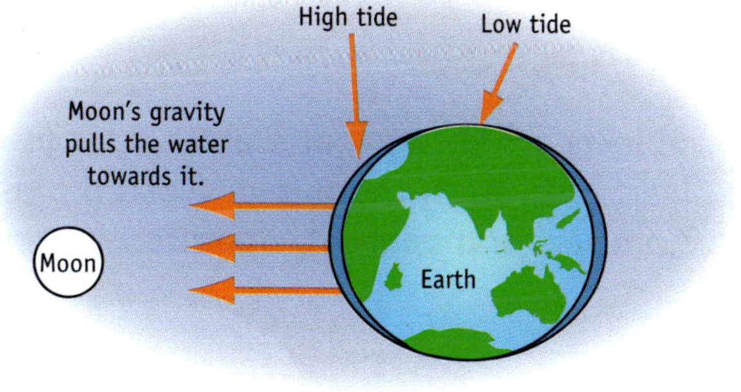

time scales

vary. Changes and events on *Earth* occur on a different time scale from those in *space*. Human time is very short when compared with geological and astronomical time.

Human time: Events related to humans involve minutes, hours, weeks and years.

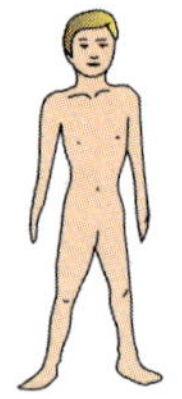

Humans have been on Earth for about 80 000 years.

Geological time: Mountains, canyons and landforms take millions of years to form.

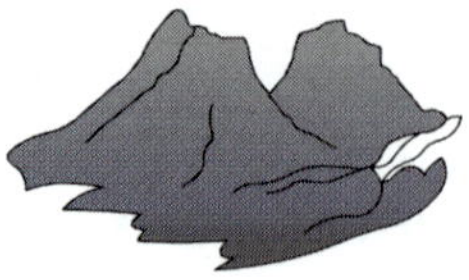

Mountains in Kosciusko area formed about 10 million years ago.

Astronomical time: Events in space take many millions of years.

It is believed the Universe began about 15 000 million years ago and our Sun is 4600 million years old.

tissues

are formed when similar *cell*s are grouped together (e.g. to form skin). The cells in a tissue have the same function (job). Different tissues are grouped together to form *organ*s.

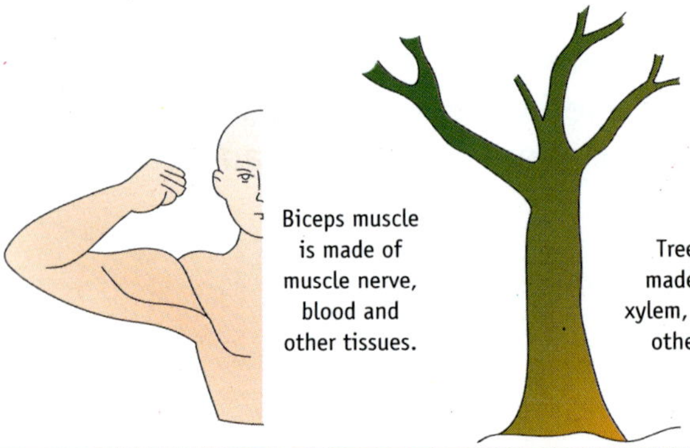

transfer of energy

See **energy transfer**.

transformation of energy

See **energy transformation**.

translucent

objects are those that scatter *light* as it passes through them (e.g. frosted glass). Compare **transparent; opaque**.

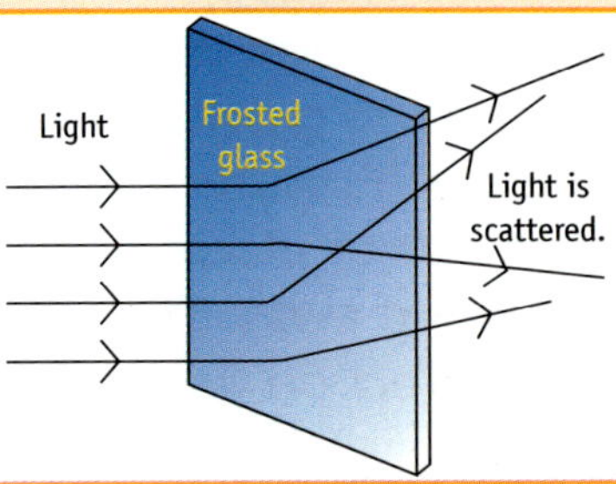

transparent

objects are those that *light* will pass straight through (e.g. glass and water). Compare **translucent; opaque**.

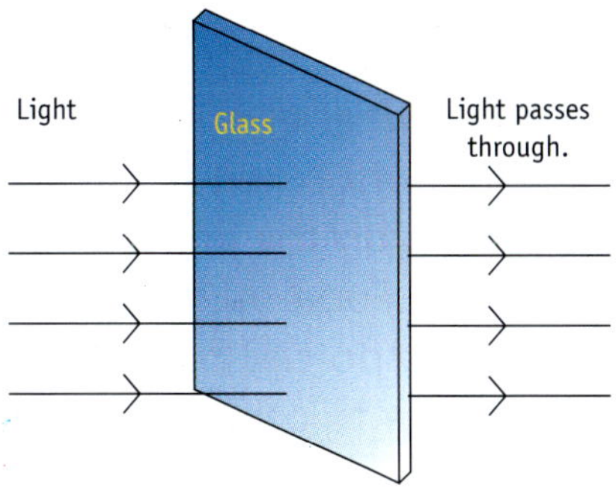

transpiration

is the loss of water from *plants* by *evaporation*. Most transpiration occurs through pores (*stomates*) in the *leaves*. When water evaporates from the pores, more water is drawn into the plant from the *soil*. *Wind* and *heat* from the *Sun* increase transpiration. See **transport systems in plants**.

Pathway of water through a plant

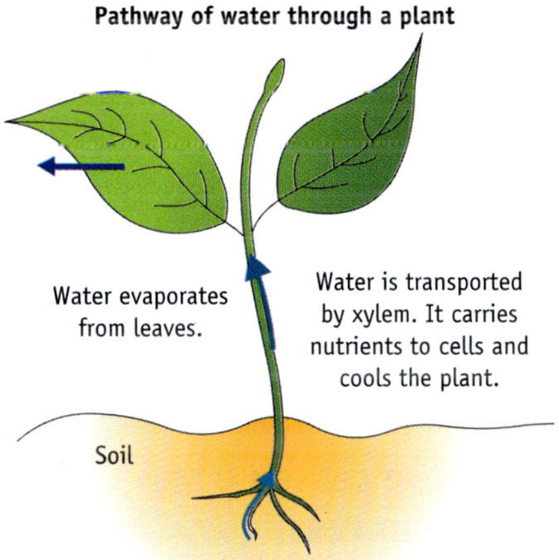

transport systems in built environments

are needed to transfer people and goods from one place to another. Cars, trucks, planes, trains and ships are part of transport systems. Roads, airports, railways and shipping ports are required for transport systems.

transport systems in plants

carry *food* through the plant. Water and *minerals* are transported or carried from the *soil* through the *plant*, in *xylem tissue*. *Food* made in *leaves* is transported by *phloem* tissue to other cells.

Pathway of water through a flowering plant

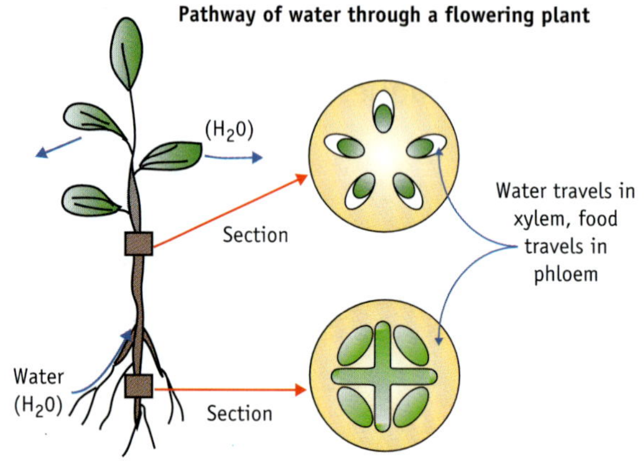

trusses

See **building and construction**.

tsunami

is a huge wave caused by an earth movement or *earthquake* in the sea floor. Tsunamis can travel at 700 km/hr and can be over 10 m high near the shore. *Seismographs* are used to detect the position of the earthquake causing the wave.

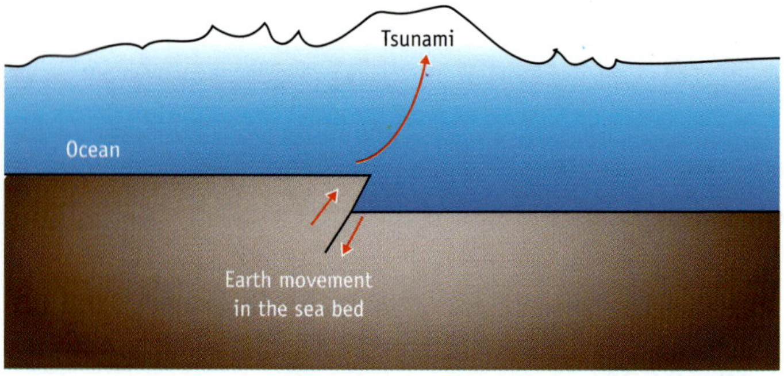

tunnel construction

can be by different methods, depending on the type of *rock* through which the *tunnel* is being built (e.g. drilling is used in soft rock, explosives in hard rock). Soft mud, underground water and cracked rocks need to be considered. Tunnels are built with ventilation and drainage and are lined with steel and concrete to prevent collapse.

tunnel types

chosen depend on the *tunnel*'s use and the surrounding *material*. See **tunnel construction**.

A bored tunnel is made by boring machines in soft rock and explosives in hard rock.

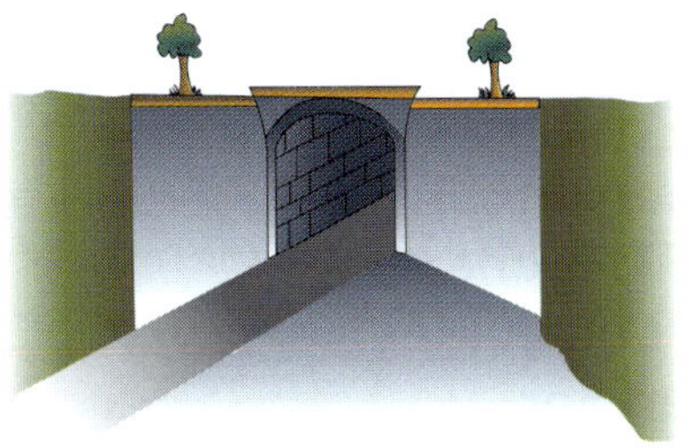

In a cut-and-cover tunnel a trench is dug, then lined and covered.

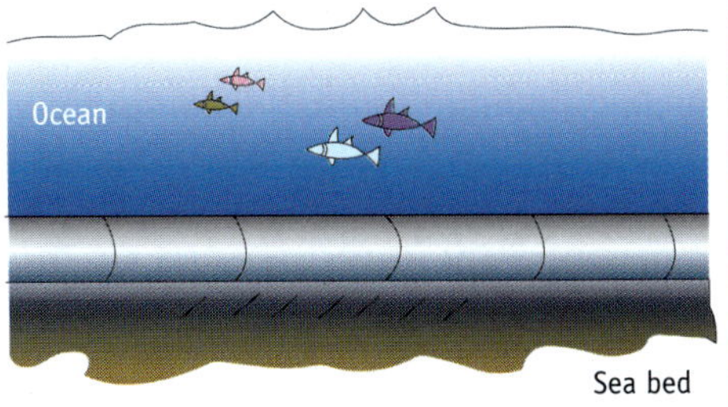

A sunken tube tunnel is used under water. Steel tubes are lowered to the sea bed and welded together.

tunnels

are made to take roads and railways underground. They may also carry sewage, water and pipelines for gas or oil. A study of the surrounding *rock* and *soil* is important when constructing tunnels. See **tunnel types; tunnel construction**.

turbidity

is the cloudiness of water caused by living and non-living particles suspended in it.

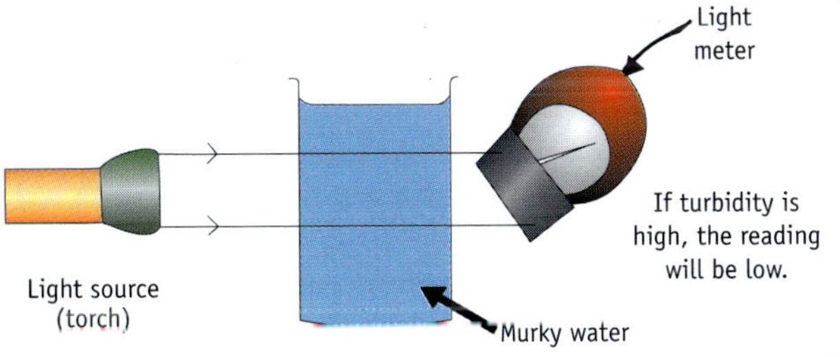

turbine

is a set of curved blades mounted on a shaft that is attached to machinery such as a *generator* or pump. *Wind*, steam or water is used to move the blades and rotate the shaft, which in turn powers the *machine*. See **windmill; waterwheel**.

ultraviolet rays

are one type of radiant energy from the *Sun*. Ultraviolet rays are also produced by fluorescent lights and arc welders. They carry more *energy* than *light* and can penetrate and damage skin (causing skin cancer) and eyes. See **ozone**.

unicellular organisms

are tiny *living things* or *microorganisms* that are made up of one *cell*. This one cell needs to be able to carry out all the *life processes* that are necessary for survival (e.g. *respiration* and obtaining *food*). Compare **multicellular organism**.

Paramecium (a microscopic animal that lives in fresh water)

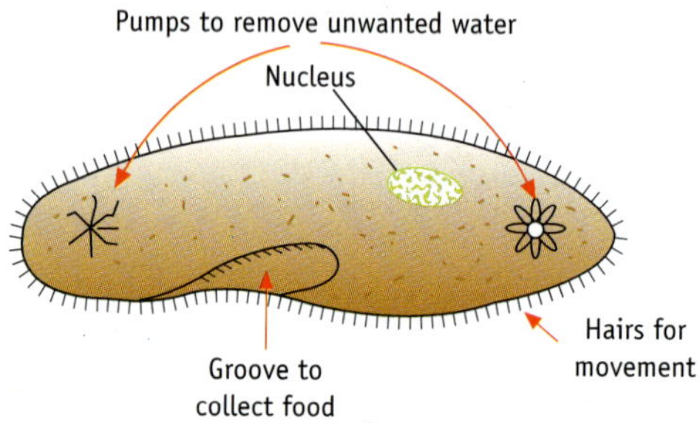

Universe

is everything that exists. It refers to all space, *matter* and *energy*. Scientists believe an explosion produced the Universe about 15 000 million years ago and it has been expanding ever since.

upthrust

is the upward *force* on an object which is in a fluid (*liquid* or *air*). It causes ships and balloons to *float* and parachutes to work and gives an object *buoyancy*. See **float**.

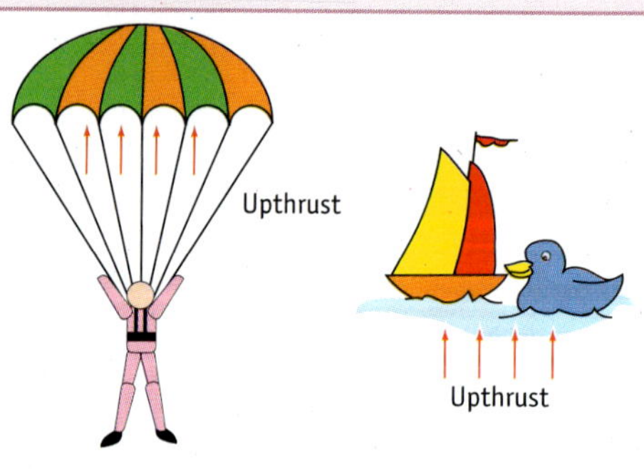

uranium (U)

is a radioactive *metal*. It is used in nuclear weapons and as a fuel in most types of nuclear reactors. *Mining* and *refining* of uranium and its use in reactors produce radioactive waste, which is difficult to dispose of.

Uranus

is one of the *planet*s furthest from the *Sun* in our *Solar System*. It consists of a *mixture* of *gas* and ice around a *solid core* and has a bluish tinge. The axis is almost horizontal. It has a system of rings and fifteen known *moon*s.

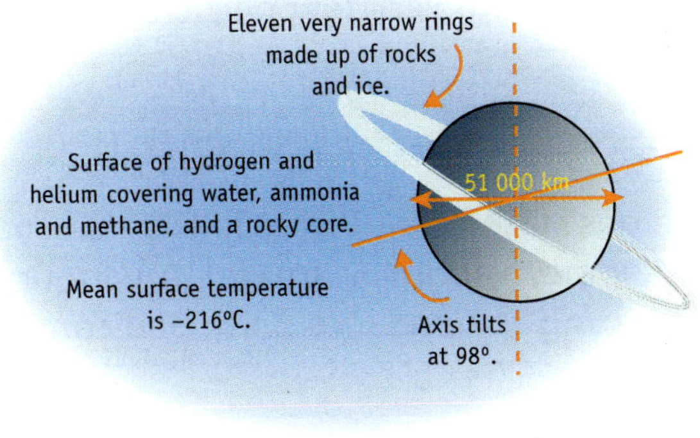

vacuum

is a space with very few particles of *matter*. A true vacuum is completely empty.

vapour

The gas state of a substance that is usually solid or liquid at room temperature (e.g. water vapour or steam). A vapour can be changed to a liquid by increasing the pressure on it. See **compression**.

variable (in an experiment or fair test)

is a factor that can change. In an experiment to see how wing shape affects the flight of a toy plane, the planes would be identical except for the wing shape. See **scientific method**.

These planes are made from the same wood and are the same size and weight. They are tested in the same conditions (wind, moisture).

variations

are the different features *organisms* possess. See **invertebrates; vertebrates; variations within a species**.

Most birds fly.

Others are flightless.

variations within a species

are the differences between individuals belonging to the same *species*. They may be produced by genes (e.g. eye and skin colour) and/or by the *environment* (e.g. body size and build caused by different diets and exercise).

Venus

(the 'morning star' and 'evening star') is a rocky *planet* with gentle hills and several active *volcano*es. The planet is covered with dense clouds made of sulphur dust and drops of *concentrated* sulphuric acid. It has no moons or water.

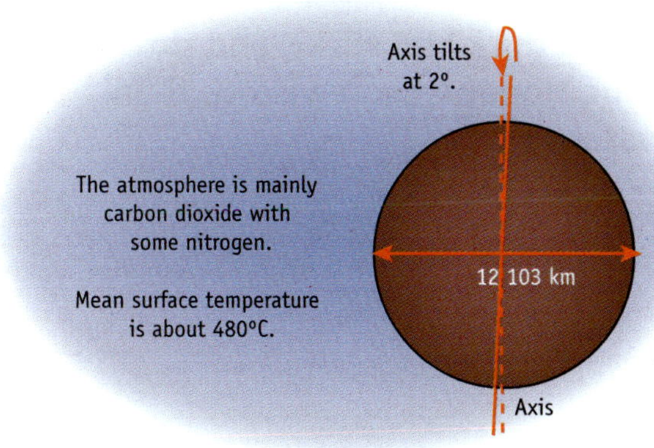

vertebrates

are *animals* that have a backbone. They are subdivided into five groups.

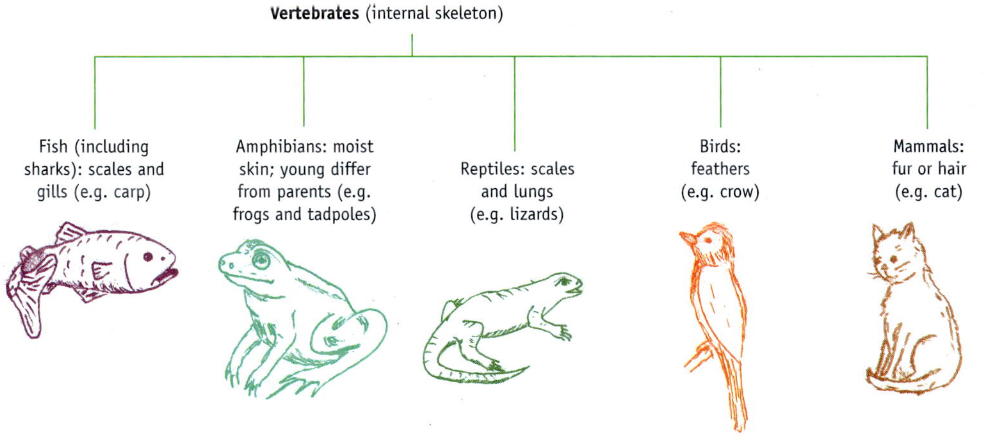

virus

is an extremely small *organism* and can be seen only with an electron microscope. Viruses are *parasite*s. They cause disease and reproduce only when inside living *cell*s.

Influenza virus

Herpes virus

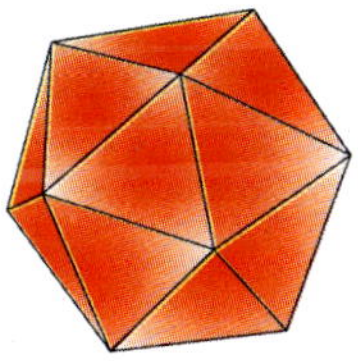
Hepatitis virus

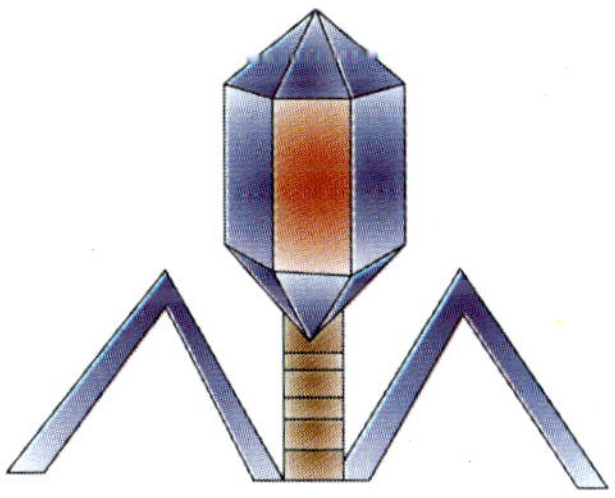
Bacteriophage virus which invades bacteria

viscous

means thick and sticky, and slow to pour. Honey is more viscous than water.

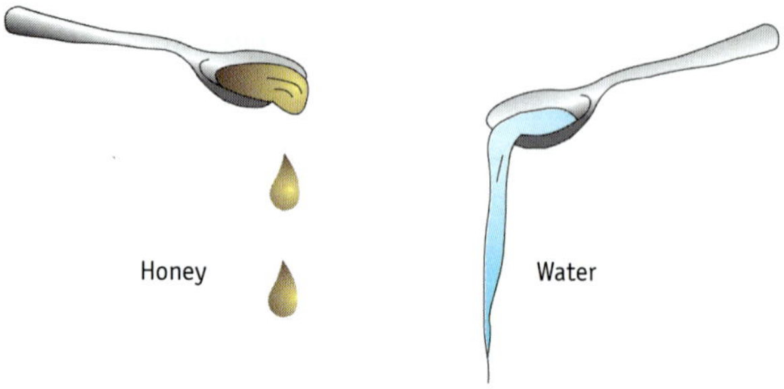

vitamins

are *nutrients* needed in small amounts for body growth, *chemical reactions* and body functioning. *Plants* make their own. *Animals* obtain most from their *food*.

volcano

is an opening in the *Earth's crust* through which *lava*, ash and *gases* flow from the inside the Earth.
See **seismograph; earthquake zones**.

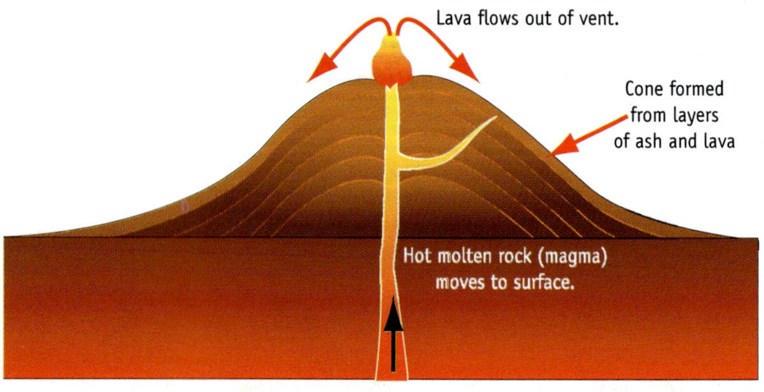

voltage

is electrical push and is measured in *volts*. It is the difference in *charge* between two parts of a *circuit* and is provided by an *electric cell*, battery or *generator*. It causes charge or *current* to move through the wires in a circuit. See **voltage in batteries**.

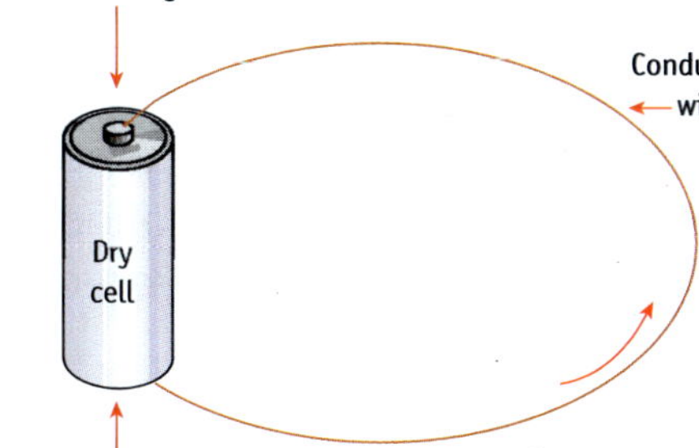

voltage in batteries

determines the strength of the battery. A battery with a high *voltage* has a high negative *charge* on its negative terminal. It is capable of 'pushing' more *current* through the *circuit* than a battery with a low voltage. See **batteries in series; batteries in parallel**.

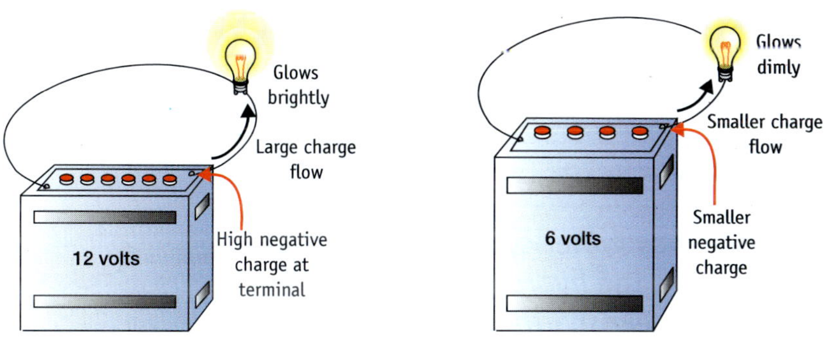

volts (V)

are a measure of electrical pressure. They are measured with a voltmeter. See **voltage**.

volume

is the space an object or substance occupies. It is measured in litres (L), millilitres (mL) or cubic centimetres (cm^3 or cc).

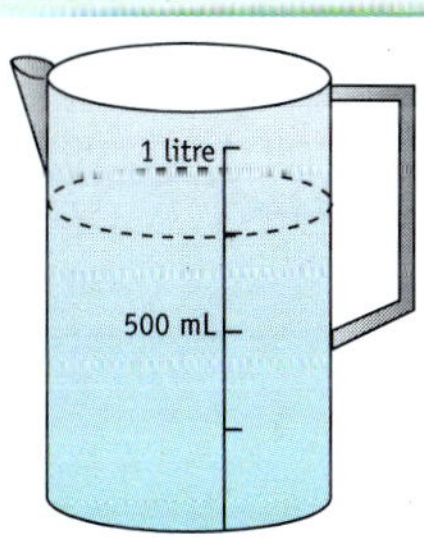

Volume of water is 750 mL

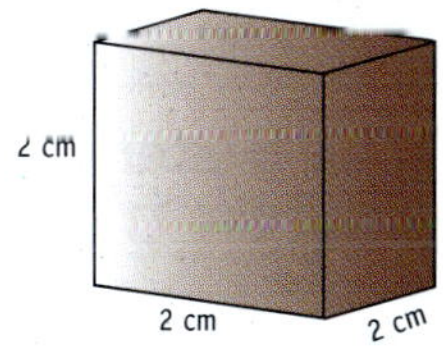

Volume of cube is 8 cm^3 or 8 cc

volume of a sound

See **loudness**.

waste disposal

in many cases consists of recycling or reusing waste *materials*. *Soluble* materials that will not pollute the waterways are often washed down the sink. *Insoluble matter* and poisonous substances (e.g. paints, pesticides, oil, *solvents*) need to be disposed of by special collection. They are treated to make them safe, and useful substances are recovered.

water cycle

is the way water cycles between the *hydrosphere* (sea, rivers, oceans), *lithosphere* (land) and *atmosphere* (*air*).

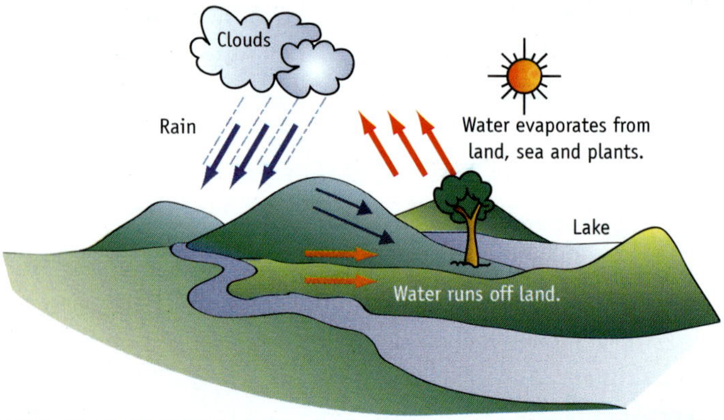

water retention

See **soils, water retention**.

water treatment

is the process of purifying town water. *Sediment*ation is the first step: the water stands in large reservoirs to allow mud and silt to settle to the bottom. The water is then taken from the upper layers and filtered and chemically treated.

water wheels

are turned by the *force* of the water on blades that are set across the rim of the wheel. The rotating wheel can be used to drive machinery. Water wheels were commonly used to drive machinery in the first factories. Compare **windmill**. See **turbine**.

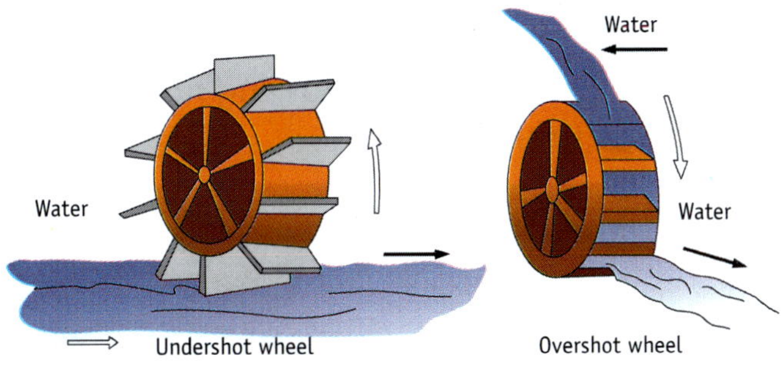

watt (W)

is a measure of *power*. The watts on the label of an appliance show how much power it draws from the *electricity* supply. An 800-watt toaster uses 800 units of *energy* every second. See **electrical energy used by an appliance**.

wave energy

is the *kinetic energy* in ocean waves and may be used to produce *electricity*. One design uses a line of floats. As they bob up and down with the water, the motion drives a *generator*, which produces electricity. See **energy sources, renewable**.

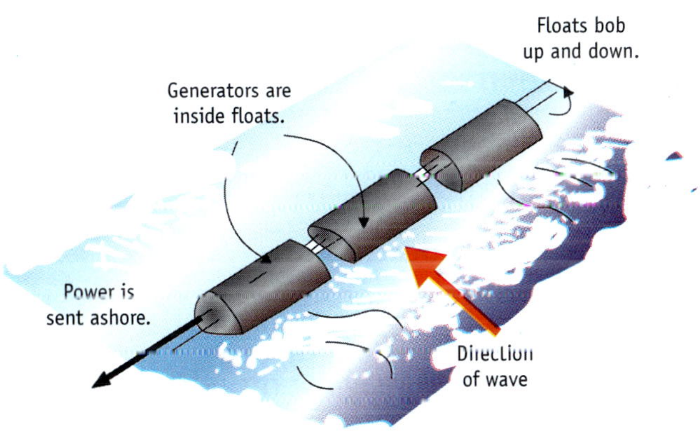

weather

is the conditions in the *atmosphere*. It includes *temperature* of the *air*, *air pressure*, *cloud* cover, *winds*, *humidity* and *rain*fall. See **oceans and weather**; **high pressure system**; **low pressure system**.

weather balloons and satellites

carry instruments that take photographs of climatic conditions and record *temperatures*, *air pressures* and *humidity* at different heights in the *atmosphere*. The information is sent by radio transmitters to *Earth*, where it is analysed, allowing meteorologists to predict the *weather*.

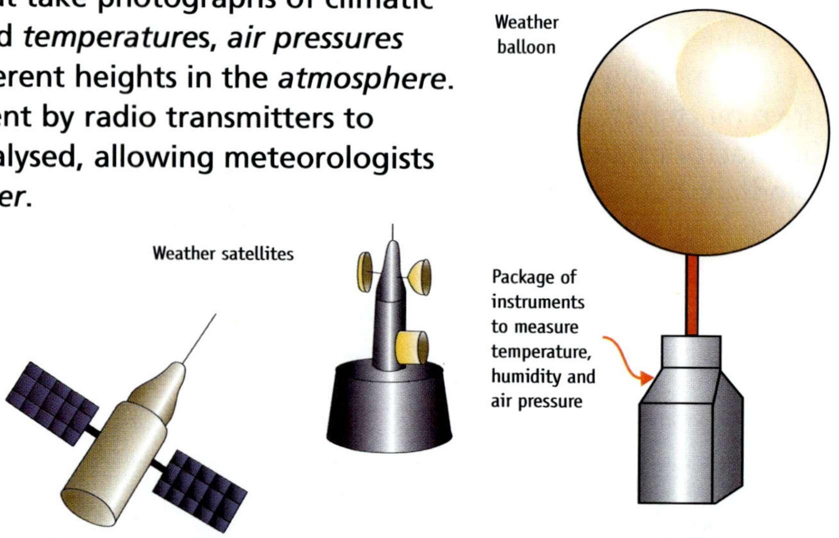

weather forecasting

is making predictions about future *weather* trends. Information about the *atmosphere* and the oceans is collected by radar, aircraft, ships, weather stations, *weather balloons and satellites*.

weathered rock

See **weathering**.

weathering

is the breaking down and crumbling of *rocks* into smaller particles. This forms *soil* or *sediments*, which may eventually form *sedimentary rocks*. See **physical weathering; chemical weathering**.

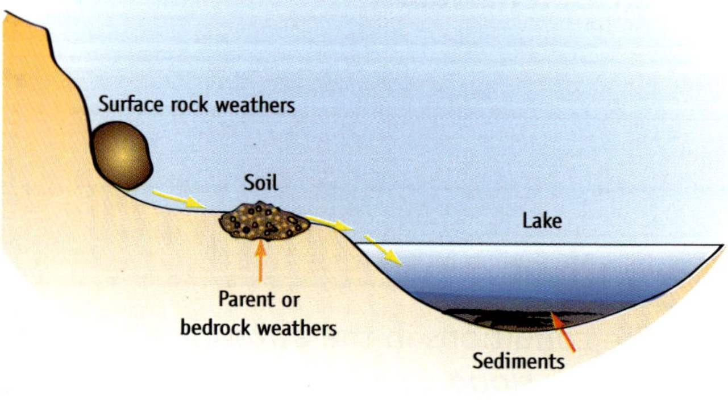

weight

is the downward *force* of an object caused by *gravity*. The *Earth*'s gravitational force is six times stronger than the *Moon*'s, so an object will weigh six times more on Earth than on the Moon.

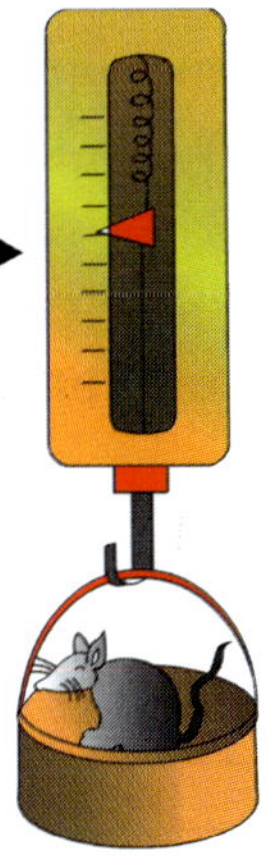

White Dwarf

is a *star* at the end of its life. It is formed from a *Red Giant*. As the Red Giant cools, the outer layers drift away and the inner layers compress, leaving a tiny *core* that shines with a white *light*. The White Dwarf shrinks away and the star dies. See **star, life span.**

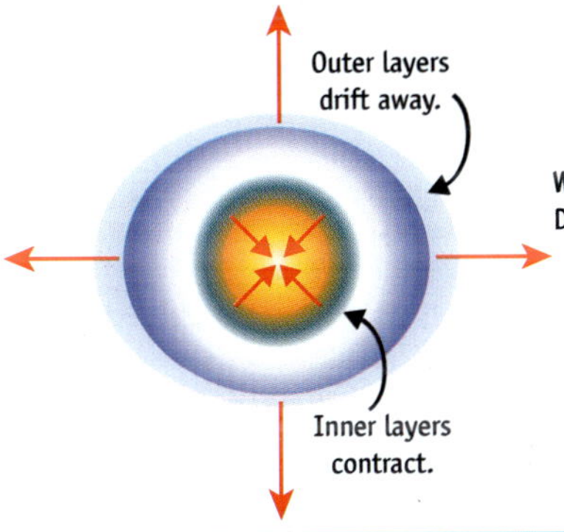

wind

is moving *air* caused by local heating. *Air* moves from areas with high *air pressure* to areas with low pressure. The *speed* of the wind is measured with an *anemometer* and the direction is measured with a *wind vane* and compass. See **high pressure system; low pressure system**.

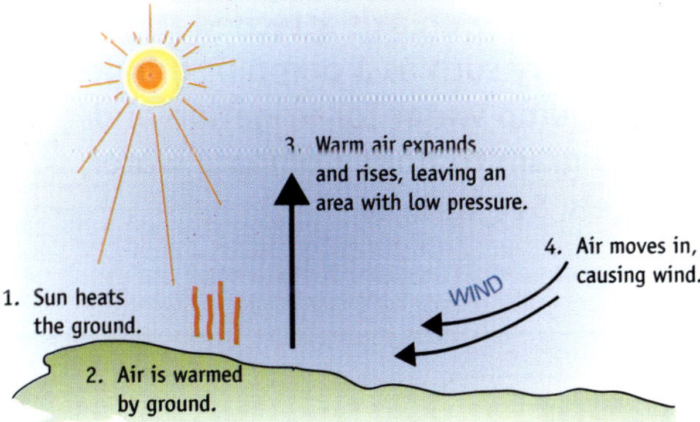

wind energy

is when wind turns propellers of *windmill*s, which can be used to pump water. This *energy* can also be channelled to a *generator*, where it is changed to *electrical energy*. See **energy sources, renewable; turbine; windmill**.

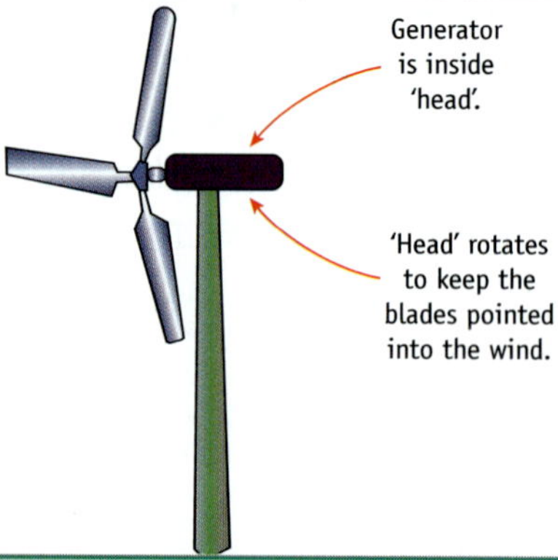

wind vane

measures the direction of the *wind*.
It indicates where the wind is coming from (e.g. a westerly wind is coming from the west).

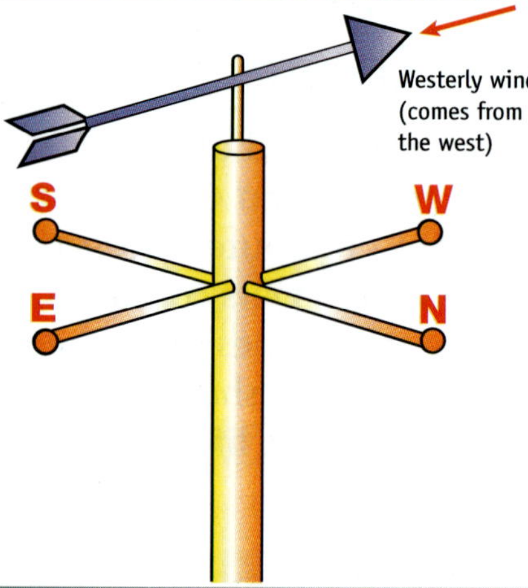

windmill

uses the *wind*'s *energy* to drive *machine*s. The wind turns angled blades, which rotate an axis attached to machinery such as a pump or grinder. They pump water for irrigation and grind grain to flour. Compare water wheels. See **turbine**.

winter

See **summer and winter; seasons**.

word equations

are used to show the chemicals that react (the reactants) and those that are produced (the products) in a *chemical reaction*. For example:

oxygen + hydrogen ⟶ water

Oxygen and hydrogen are the reactants; water is the product.

work

is done on an object when it is changed or moved by a *force*. *Energy* is transferred (moved) or transformed (changed). See **energy transfer; energy transformation.**

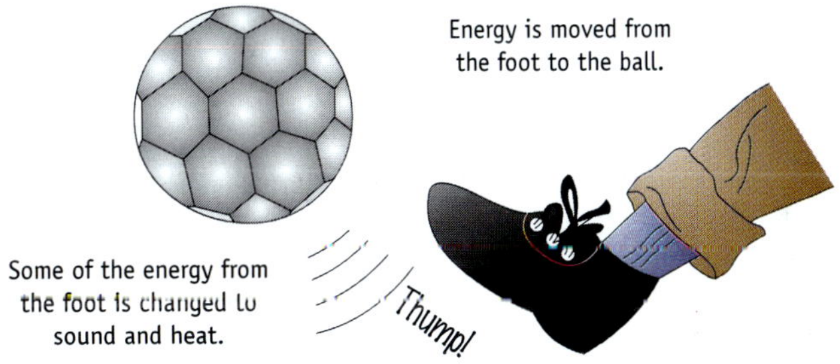

X

X-rays

are one type of radiant energy produced by the *Sun*. Humans produce them with an X-ray tube. They readily penetrate many materials and are useful in *science*, medicine and industry. For instance, they are used to see inside the body. Bones and teeth stop the X-rays and they appear as a shadow on the photographic plate.

xylem

tissues carry water in plants. The *cells* in this tissue are dead and hollow and function like microscopic pipes. See **transport systems in plants.**

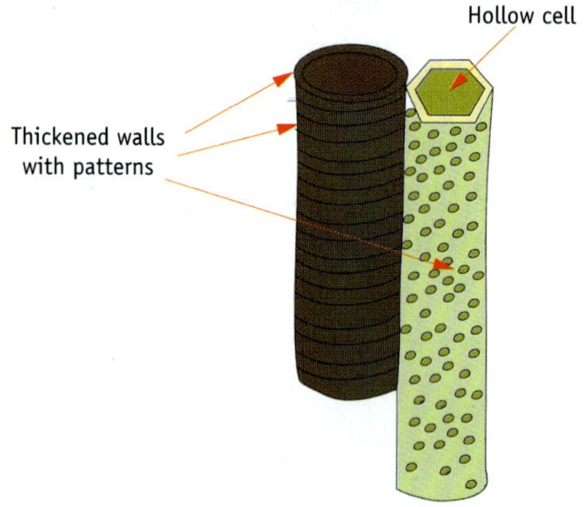

year

is the time taken for a *planet* to travel around the *Sun*. *Earth* takes 365 *days* to *orbit* the Sun. *Mercury* takes 88 *days* and has the shortest year of the *planets*. *Pluto* takes 248 Earth years and has the longest year.

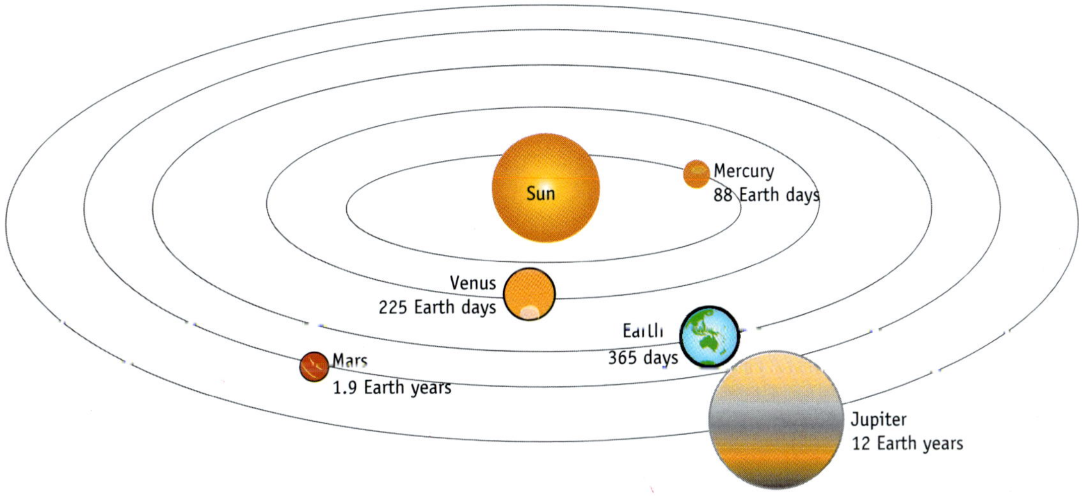